I0560729

The Heavens, the Waters, and the Partridge

The Heavens, the Waters, and the Partridge

The Historical Interaction of Faith and Science Before Modern Science

Winston Ewert

Description

The history of the interaction between Christianity and science is often treated as though it began with modern science. However, Christians have grappled with the interaction of their faith and the science of their day since the days of the early Church. *The Heavens, the Waters, and the Partridge* explores this fascinating and underappreciated history. This book explores various issues, looking at the Scriptural and scientific background, how different historical Christians interacted with the issue, and how that interaction has fared in the light of modern science. This history provides lessons that will help Christians know how to think about the interaction of Christianity and science today.

Copyright Notice

©2025 by Winston Ewert. All Rights Reserved.

Library Cataloging Data

The Heavens, the Waters, and the Partridge by Winston Ewert
496 pages, 6 by 9 inches
Library of Congress Control Number: 2025933556
ISBN: 979-8-9917040-4-5 (Hardcover), 979-8-9917040-6-9 (Paperback), 979-8-9917040-5-2 (Kindle)
BISAC: REL067070 RELIGION / Christian Theology / Ethics & Moral Teaching
BISAC: SCI034000 SCIENCE / History

Book Cover Design:

The image on the book cover was generated with AI.

Publisher Information

Inkwell Press
2321 Sir Barton Way
Suite 140-1032
Lexington, KY 40509
Inkwell.Net

Advance Praise for *The Heavens, the Waters, and the Partridge*

Christians have been wrestling with the relationship between science and faith since the early days of the church. Yet surprisingly, there has not been an in-depth study of how the church fathers engaged classical science with focus on what we can learn for the current debate. *The Heavens, the Waters, and the Partridge* addresses some of the most interesting science-faith questions such as the existence of the Star of Bethlehem, the size of the Ark, and *creatio ex nihilo*. You will both learn from and enjoy this book.

—Sean McDowell, Ph.D.

Author, Speaker, Biola

The Heavens, the Waters, and the Partridge fills a much-needed gap in the study of Christianity's relationship with science. In it, Ewert covers 28 scientific controversies—from the nature of the heavens to the bird in the title—with which famous Christian writers from the past grappled. Sampling Christian luminaries such as Basil, Augustine, Aquinas, Calvin, and Luther, Ewert shows that the Church has always faced attacks from those who thought science and Scripture were at odds. More importantly, by showing us how those writers dealt with the controversies of **their** day, he helps us to better understand how modern Christians should deal with the controversies of our day. This book is a valuable resource for anyone who wants to fully understand the relationship between Christianity and science.

—Jay L. Wile, Ph.D.

Adjunct Professor, Memoria College

Winston Ewert's carefully-documented and clearly-written book fills a huge gap in the study of the relationship of Christianity and science. While most treatments pick up the story of this contested relationship after Copernicus (and especially after Darwin), Ewert addresses that history before Copernicus, and what he finds is of great value to historians, scientists, philosophers, and Christian apologists. As he says, "If we get the history of the interaction of Christianity and science wrong, it will be very difficult to get that same interaction right in our modern day. If we do not know the history, we cannot learn the lessons of the past and make an informed decision about how to act today."

—Douglas Groothuis, Ph.D.,
Distinguished University Research Professor
of Apologetics and Christian Worldview,
Cornerstone University and Seminary.

To Elyse,

Asking you out significantly delayed this book,
and I don't regret that at all.

Table of Contents

Foreword

C.S. Lewis lamented that the reason people do not believe the Bible is simply because it is old.[1] For the thoughtful person, this should never be grounds for disbelief, and since you are reading a book subtitled *The Historical Interaction of Faith and Science Before Modern Science*, you probably do not fall prey to the erroneous equation: old = false. I'm ashamed to admit that I subscribed to this wrongheaded view while a first-year undergraduate at MIT. I was, therefore, fortunate that year to receive a copy of Hugh Ross's excellent book *The Fingerprint of God*. Published shortly before I started college, it described the latest cosmological evidence and how this evidence supported a belief in theism. As a result, my faith in God, the Bible, and the Judeo-Christian worldview was strengthened. I saw how these "old" religious concepts could be supported by newer and therefore (in my mind!) truer realities.

Looking back, I would have been well served if the book you now hold—or have open on your e-reading screen—had also been published in 1989 and had supplemented my college reading list. Dr. Winston Ewert, a renowned computer scientist and thinker, has written an antidote to the small-mindedness that views modernity, with its flashy technology and new-car

---

[1] The definition of "chronological snobbery" is "the uncritical acceptance of the intellectual climate common to our own age and the assumption that whatever has gone out of date is on that account discredited" Lewis, *Surprised by Joy*, 207–8.

smell, as the sole proprietor of all the answers about science, faith, and their interdependence. The helpful impact of Ross and other authors notwithstanding, we would do well to heed Ewert's take-home message: Christians need a healthy dose of humility before attempting to reconcile the Bible with science.

Ewert supplies this dose of humility, first, by showing how early Christian theologians have attempted to reconcile science with the Bible many times and, second, by demonstrating how these attempts sometimes end badly. Surprisingly, Ewert shows how these attempts go badly, not because the biblical account is so scientifically incorrect that reconciliation was doomed from the start, but because two things repeatedly derailed these attempts: bad science and bad biblical understanding.

The encouraging thread woven through this book is that when Christian scholars returned to the true sources of biblical and scientific understanding, they avoided these two pitfalls and admirably connected the Bible to the natural world. For example, nearly all the Christian writers who tackled the concept of spontaneous generation wisely went back to the scriptural source and saw that the Bible does not teach this concept—which we now know to be false. By returning to the source, they avoided the trap of trying to read into the Bible things that the Bible did not say, which would have been tempting to do, considering that all of their contemporaries believed and taught this false theory. Moreover, happily, there were no mistranslations of key passages that, for other topics, confused the issues. Also, the ancient Christian thinkers went to the scientific sources and realized that their pagan contemporaries did not have rock-solid evidence and, thus, knew that conjectures about spontaneous generation were open to debate. The combination of not misreading the Bible, not having poor biblical translations, and not buying into unsupported scientific theories served them well. Air-tight (quite literally, in the case of Louis Pasteur!) experiments later confirmed that these early Christians were more correct than their secular counterparts.

As you will see in this book's many chapters, early Christians did not always exhibit the universal agreement they had in their rejection of spontaneous generation and other now-debunked theories. However, as Ewert artfully explains, the writers who turned out to be correct—or at least less incorrect relative to their secular peers—were the ones who avoided bad science and bad biblical understanding.

The way Ewert instills appreciation for these ancients, both pagan and religious, reminded me of an experience I had co-teaching a course in 2005, *The Interplay between Science and Religion in Western Culture*, with my wonderful Wake Forest physics colleague and respected cosmologist, Dr. Paul Anderson. As we wove together our distinct visions for the class, Paul introduced me to the amazing ways Greek astronomers worked out the size of the Earth, the relative sizes of the celestial bodies, and even the distances to the sun and the moon.

Likewise, after reading Ewert's book, one begins to admire the ancients' reasoning powers. Despite their difficulties in measuring the size of the heavenly bodies, the nature of the Earth's core, and other topics that their limited technology did not allow them to explore, pagan and Christian thinkers were still able to tease out many of the ways our universe works. Our generation no longer has these technological constraints, but are we reasoning as carefully as our predecessors?

I found that by reading Ewert's book my reasoning and my faith were both strengthened. This did not surprise me. I have known Dr. Ewert for more than a decade, and I have often seen firsthand his ability to delve deeply into complex topics to forge something of exceptional value—something that causes people to think more clearly. Most recently, I witnessed Ewert compile every variant genetic code known to modern science. No small feat! His published manuscript on the codes is a treasure trove, even for scientists who vehemently oppose Ewert's conclusions. Researchers of all stripes drink from Ewert's well of knowledge for the simple reason that no one else has been able to draw together such a vast amount of information and present it in a clear, elegant format.

Ewert embarked on his heroic compilation of variant codon tables for much the same reason he wrote this book: because he was curious. As he documents in this book's preface, Ewert's starting point for this book was an oft-quoted passage from Augustine. In that quote, Augustine bemoans Christians speaking nonsense and attributing it to the scriptures. But Ewert wondered, what "nonsense" did Augustine have in mind? From there, he followed the rabbit hole down, down, down to a world of deeply intelligent, deeply spiritual thinkers who sparred with their secular peers over the biggest questions of their day. As he journeyed through the historical writings, Ewert drew together an enormous body of facts concerning what theologians of the early church wrote about various scientific issues. There are many experts in church history, and many who specialize in various spiritual luminaries. But Ewert accomplished something that no one else had yet attempted: he distilled their intellectual engagements with respect to the natural world and presented these interactions in a clear, elegant format.

This book could not be more timely. We need people of faith who use careful reasoning and avoid the pitfalls that surround us today and which, as Ewert shows, have always been a threat to sound thinking. By turning to the past, Ewert has unlocked opportunities for us in the present—opportunities to reexamine our pet theories, the prevailing wisdom, and the questions that have haunted humanity from time immemorial.

Thankfully, this process of delving into the past is quite enjoyable. The various well-known players (Augustine, Aquinas, Dante, Luther, and Calvin), the authors we wish we knew better (e.g. Justin Martyr, Chrysostom, Jerome, Origen, and Athanasius), and many we may never heard of (Cosmas Indicopleustes, Remigius of Auxerre, Hilary of Poitiers, and at least a dozen more) all march across Ewert's stage defending their various theories about the way the world works. It is extremely enjoyable reading. If ever you have wanted to know more about early Christian writers, this is as painless a route as can be imagined. The topics they discuss are fascinating. Who doesn't wonder about the Star of Bethlehem or the long lifespans recorded in the early Genesis genealogies?

The topic of patriarchal longevity has special significance. Phillip E. Johnson, wildly regarded as the "godfather" of Ewert's intelligent design community, recommended that Christians start with *that* controversy, which is recorded in Genesis chapter 5, about the longevity of the patriarchs, before taking on more contentious issues such as the age of the Earth, which is the focus of Genesis 1:[2]

> I am not opposed to discussing the days of Genesis 1 or any other topic, but I think most of us will be better prepared to do so after we have had the opportunity to think through the full implications of what we conclude about Genesis 5.

Unknowingly (he told me that Johnson's suggestion from *The Right Questions* was <u>not</u> his inspiration for chapter 25), Ewert has taken Johnson's advice and done us one better. He has not only shifted from the highly contentious Age of the Earth issue to issues that provoke more curiosity and less dismissiveness ("Did the Antediluvians Live for Centuries?" is just one of several such chapters). But Ewert has also taken us back in time.

---

[2] Johnson, *The Right Questions*, 147.

As he puts it in the introduction, this time shift has a distinct advantage: "We can be observers rather than participants in these debates." What a brilliant maneuver to generate less heat and more light!

Plus, it is pure pleasure to read about the questions that captivated ancient minds. It takes us back to the eras when one did not need to own a multimillion-dollar laboratory to weigh in on big scientific questions—*Does the moon have its own light? Are the heavenly bodies alive? Do stars determine our actions?*—and many other questions that, like these three, are chapter titles in Ewert's book. Simply by observing the world around them and carefully examining the scripture, early Christians were able to go toe-to-toe on these questions against the best that the secular world had to offer, and they often came out on top ("No, Celsus dear, the stars do *not* determine our actions, thank you!")

As Ewert concludes in his final chapter, we should take heart in learning how our predecessors approached these questions. Their example is far from being false simply because it is old; instead, it should inspire us to grapple tenaciously with the issues of our day. As we do, we must avoid the trap of reading into the Bible what we want it to say. This is especially important if one is attempting to reconcile everything in the Bible with everything in modern science. Perhaps, as was the case in the past, not everything in modern science will turn out to be true. A careful handling of scripture combined with a careful handling of received scientific wisdom will serve us well. Following Ewert's excellent advice, we should always go back to the source!

Jed Macosko
May 22, 2025
Winston Salem, NC

Preface

Our subject in this book is the history of the interaction of Christian theologians and apologists with the science of their day before the advent of modern science. For our purposes, we take modern science to begin with the Copernican Revolution which began in the sixteenth century. Much has been written about the interaction of the Church with the science of the modern era, but relatively little focus has been placed on that interaction before the modern era.

I would seem like an odd choice for the author of such a work because I'm neither a historian nor a theologian, but rather a computer scientist. For computer scientists, nothing worth mentioning happened in history before 1900, let alone 1543. How did it arise, then, that I took upon myself the task of writing this book?

I have for some time been involved in the debate over intelligent design and evolution. I have used my skills as a computer scientist to defend the traditional Christian claim that the world was designed by a supreme intelligence and that we can discern this from our study of the natural world. Intelligent design has many critics. Some of these critics are non-Christians who are understandably hostile to any claim that the Christian God or anyone like him is responsible for the creation of the world.

However, other critics come from within the faith. These Christian critics still believe that God created the world, but believe that he acted in a way compatible with mainstream scientific theories about its origin and development. For them, intelligent design is not just wrong; but dangerous. They believe that the Church must make peace with the science of our day, accepting what its adherents take to be well-established scientific theories. Opposing these theories can have no other effect than to expose the gospel and our faith to ridicule.

These critics sometimes quote the immensely influential Church father, Augustine, in support of their position:[1]

> Now, it is a disgraceful and dangerous thing for an infidel to hear a Christian, presumably giving the meaning of Holy Scripture, talking nonsense on these topics; and we should take all means to prevent such an embarrassing situation, in which people show up vast ignorance in a Christian and laugh it to scorn. The shame is not so much that an ignorant individual is derided, but that people outside the household of faith think our sacred writers held such opinions, and, to the great loss of those for whose salvation we toil, the writers of our Scripture are criticized and rejected as unlearned men. If they find a Christian mistaken in a field which they themselves know well and hear him maintaining his foolish opinions about our books, how are they going to believe those books in matters concerning the resurrection of the dead, the hope of eternal life, and the kingdom of heaven, when they think their pages are full of falsehoods on facts which they themselves have learnt from experience and the light of reason?

Augustine writes of Christians who defend ignorant nonsense and attribute that nonsense to the Scriptures. They believe that their position is taught in Scripture, but reason and experience

---

[1] Augustine, *The Literal Meaning of Genesis. 1*, Book 1, Chapter 19.

show their ideas are undeniably false. As such, they are quickly exposed as ignorant fools. This would not be so bad if it were simply a case of ignorant men being subjected to ridicule. However, because they have invoked the Scriptures in support of their ignorance, now the Scriptures are also subject to the same derision. Those outside of the faith, seeing this, will be led to conclude that the Scriptures also teach this ignorant nonsense. As a consequence, they will reject the whole of Scripture and their souls will be lost.

Certainly, nobody could reasonably dispute the central point that Augustine is making. No Christian wants to needlessly introduce stumbling blocks to those outside the faith by incorrectly attributing false views to the Scriptural text. All Christians can profitably learn a lesson of care and discernment in what views they hold and, especially, which views they claim are taught in Scripture.

However, nobody thinks this about their own views, but rather about the views of other Christians. Theistic evolutionists think this about old-earth creationists. Old-earth creationists think this about young-earth creationists. Young-earth creationists think this about geocentrists. Geocentrists think this about flat-earthers. Yes, there are today a small number of Christian geocentrists[2] and flat-earthers.[3] Nobody thinks of themselves as the problem, but rather that those other Christians are the problem. Thus, the holders of these various positions will argue among themselves about which views are ignorant nonsense, needlessly introducing stumbling blocks to the faith, and which are actually taught by the Scriptures.

---

[2] Sungenis, Robert and Bennet, Robert, *Galileo Was Wrong: The Church Was Right: The Evidence from Modern Science.*

[3] Hart, Pauly, *Biblical Cosmology.*

However, when I read this quotation, I found myself asking a somewhat different question: What in the world was Augustine talking about?

Augustine lived in the fourth and fifth centuries, long before the advent of modern science. The controversies that we have today over Darwinian evolution, universal common ancestry, and the age of the earth were still far into the future. Augustine could not have had our modern controversies in mind when he wrote that quotation. But what did he have in mind? What were the foolish, ignorant claims that Augustine was concerned about Christians making? I realized that I had no idea what issues Augustine was concerned about.

I dislike not knowing something, so I set about to find out by reading Augustine's book, *On the Literal Meaning of Genesis*. To my surprise, Augustine does not merely briefly touch upon science. Instead, the interaction between the creation narrative and the science of his day is a major theme of Augustine's work. Where I expected Augustine to primarily develop symbolism found in the Genesis text, he instead primarily evaluates and attempts to resolve various difficulties in the text. Some of these difficulties were scientific, whereas others were more philosophical or theological. Overall, his commentary primarily focuses on resolving various objections that were raised against the creation account. This was not what I had expected, but I found it very interesting.

So began my journey into the history of the interaction of Christianity with the science of its day. Augustine, I quickly discovered, was not alone. Other Christians throughout history have also interacted with the science of their own day. Examples include early church fathers such as Basil and Chrysostom, scholastics such as Aquinas, and reformers such as Luther and Calvin. They discussed a surprising variety of issues. I began to understand that there was a rich history of the interaction between the Church and pre-modern science, but not one that is widely known or appreciated. It is a story that I think deserves to be better known and appreciated.

At the same time, I increasingly understood the importance of historical narratives. Each camp in these debates has its own historical narrative, a telling of history in a way that makes its own position obviously correct. This can take the form of a narrative in which prominent historical figures agreed with its position or else as interpreting past figures who did not agree as cautionary tales whose mistakes we should avoid. In either case, the same conclusion can be drawn: history unreservedly supports the position of the narrator who happens to be telling that history.

For example, those hostile to Christianity invoke a warfare narrative in which the Church and science have always been at war. In this narrative, the Church rejected the enlightened science of the ancient Greeks and plunged the European world into the Dark Ages. When modern science belatedly arrived, the Church, in a series of skirmishes, was forced over and over again to acquiesce to science and give up part of its teaching. Our modern debates are simply the latest of such skirmishes, which the Church is certain to lose. Those who can really see what is going on simply reject the Church and put their faith in the repeated victor, science.

On the other hand, theistic evolutionist Christians, who advocate accepting modern scientific theories, including those controversial within the Church, present a very different narrative. The Church accepted the classical science of the Greeks, tweaking their interpretation of Scripture to accommodate it. There were some who resisted that science, but these were in the minority and now serve as cautionary tales of what not to do. When modern science arrived, the Church, with some missteps and delays, accepted its findings while adjusting their understanding of Scripture where required. Throughout history, the Church has recognized the expertise of scholars and scientists and sought to show the compatibility of the science of the day with the faith. Modern rejection of an old earth and evolution are thus out of step with the historical practice of the Church. If we followed the practice of the historical Church, we would simply accept these theories and adjust our reading of the Scriptures to align with them.

Another narrative is often promulgated by creationists, who reject at least some modern scientific theories that they see as in conflict with the Scriptures. In this narrative, there has historically been minimal conflict between the Church and science. There were some mistakes made, such as in the Galileo affair, but for the most part, Christianity and science have been in easy harmony. However, in the modern age, science has become corrupted to serve an anti-Christian agenda, rejecting God's role in creation. True science is still in accord with the Bible, but Christians must oppose the false science proceeding from a corrupt scientific establishment. Slight variations of this narrative are promulgated by flat-earthers or geocentrists, who simply differ as to the time when science became corrupt.

Each of these narratives can point to certain people and events in support of their story. There is some truth to each of them. However, each also trades to some extent in misconceptions and misunderstandings of those events. Furthermore, each ignores or at least downplays the events that do not so neatly fit into their story. The true history is more complex, more nuanced, and more worth knowing.

Furthermore, the true history is not merely a matter of intellectual curiosity. False narratives lend support to false views. If we get the history of the interaction of Christianity and science wrong, it will be very difficult to get that same interaction right in our modern day. If we do not know the history, we cannot learn the lessons of the past and make an informed decision about how to act today.

In this book, I seek to share what I have learned about the history of the Christianity-science interaction. I do my utmost not to simply present those parts that would support my agenda, but instead to share the whole story, warts and all. This is not to pretend that it is possible for any author to present a purely objective presentation of history, but I try to at least be fair. As much as I would like you to adopt my views on these issues, it is not my intent to convince you to do so. Rather, I want to instill an appreciation for the rich history that the Church has in this area. I want to help ensure that those who think about this subject have an accurate understanding of that history. Only once that is in place can we begin to properly consider how we should interact with these issues today.

Chapter 1

Introduction

Why?

Modern science is typically taken to have begun with the Copernican Revolution, which saw the rejection of geocentrism in favor of heliocentrism. We will call the set of ideas that prevailed before this classical science. These ideas primarily trace back to the ancient Greek philosophers, but during the Scientific Revolution, they were either rejected or changed beyond recognition.

What can we learn from how historical Christians who lived in the ancient and medieval worlds interacted with this classical science? For many of us living in the modern age, the answer may seem obvious: nothing at all. In the common conception, those living before the rise of modern science held to a mishmash of old wives' tales, superstitions, and baseless speculations. Whatever knowledge they might have had about the world, it was not science. Science is a distinctively modern phenomenon, or so we moderns tend to think. As such, it may seem doubtful that we can learn anything from our spiritual forefathers when it comes to science. However, this mindset is rooted in some misconceptions that need to be resolved.

The first misconception is that the pre-modern world was completely scientifically ignorant. For example, many moderns are under the impression that everyone in the ancient world believed that the earth was flat. But this is a misconception. The ancient Greek and Roman world not only knew that the earth was spherical, but they also had a good idea of its size. They made great strides in fields such as mathematics, medicine, and astronomy. They believed many things that would turn out to be incorrect, but they were not utterly ignorant.

The second misconception is that their theories were simply baseless speculations. Instead, their ideas were based on experience and reason. They observed the natural world and tried to make sense of what they saw. They came up with sophisticated arguments for particular explanations. They considered and debated different theories, choosing the ones that seemed best. One can certainly find fault with aspects of their methodologies. This is because the methods of science have been refined over time and are continuing to be refined. However, it would be a great mistake to dismiss their science as baseless speculation.

While we know that the ideas of classical science were flawed and would be overturned, the Christians living at that time did not. For them, classical science was science. These were the ideas deemed correct by centuries of scholars who had carefully considered them. Disputing those ideas would have seemed akin to challenging the claims of modern science today. They could no more easily dismiss classical science as baseless than we could dismiss its modern counterpart as baseless.

This matters because scientific objections to Christianity are not a modern phenomenon. Just as the Bible makes claims that are argued to be at odds with the findings of modern science, so too does it make claims that were argued to be at odds with classical science. Since its inception, Christianity has had to deal with these sorts of issues. Some of these are the same issues confronting us today, while others are different. The Church has a long history of interaction with science.

But why should we care about what historical Christians had to say about obsolete science? While the conclusions of classical science were not baseless, they were seriously flawed. What can possibly be learned from how Christians interacted with this flawed science? While the scientific framework has changed, our faith has not. We accept the same Scriptures and believe the same essentials of doctrine. While the particular conclusions they drew may no longer be relevant, we can learn from how they reached those conclusions.

In fact, studying obsolete scientific issues has an advantage. We can be observers rather than participants in these debates. This allows us to see the debates and issues more clearly. Our view of these debates is, hopefully, less clouded by our modern agendas.

Another reason to study these historical Christians' responses is that they are, generally, not partisans of our theological debates. These influential theologians have stood the test of time and predate our modern divisions and debates. They are the common heritage of most branches of modern Christianity. Consequently, we can learn lessons from them for all Christians and not merely for the particular branches of Christianity.

An additional benefit is the longevity of classical science. These ideas held sway for more than a millennium. There were developments along the way, but the basic structure remained in place from the early Church until the Reformation. As such, we can see how a variety of individuals interacted with this science across a thousand years and more.

Appealing to Our Historical Forefathers

However, it is very easy to appeal to our spiritual forefathers in a misguided and unhelpful fashion. A common tendency is to read these authors as though they were commenting on our modern debates. However, they were commenting on the issues of their own day, not ours. When it comes to questions of science, the issues were different in the era before modern science. There may be some similarities in the issues, but the whole scientific context has been radically altered. As such, to simply read our forefathers as commenting on modern issues rips them completely and inappropriately out of context.

For example, young-earth creationists like to appeal to the universal agreement among the Fathers that the world was created in the past several thousand years. Even Fathers who did not understand the creation days as ordinary-length days still affirmed the recent creation of the world. However, these Fathers were not aware of modern scientific theories or modern evidence for an ancient world. As such, critics of the young-earth position argue that these same Fathers would have come to different conclusions today. After all, even the most convinced young-earth creationist has accepted modern science on issues such as heliocentrism, spontaneous generation, and atomic theory. Thus, these critics suggest that these Fathers would have accepted an ancient world had they lived today.

Unsurprisingly, many Christians believe that the Church Fathers would have come to whatever position they themselves hold. The young-earth creationist argues that they would have maintained their belief in a young world. The old-earth creationist argues that they would have reconciled an old earth with the Biblical text while rejecting evolutionary theory. The theistic evolutionist argues that they would have accepted both an ancient world and evolutionary theory. Thus, the issue becomes embroiled in an intractable argument about what position historical Christians would have taken had they lived today. Very little is accomplished by such appeals.

If we want to benefit from the insights of our spiritual fore-fathers, we need to stop being so selfish. We cannot begin by asking what these Fathers thought about *our* scientific issues. We need to begin by asking what they thought about *their* scientific issues. If we want their help in dealing with our thorny scientific issues, we need to study how they dealt with their thorny scientific issues. We should not ask about questions that are controversial today; we should ask about questions that were controversial in their day. We need to understand the issues of their day and how they dealt with those issues, and only then can we draw insights about our own issues.

The Bible and Science

The Bible, in most English translations, does not use the word "science." The word "science" has only had its current meaning in English since the middle of the nineteenth century. Prior to that, *it referred to scholarship more generally,* and still does in other languages such French and German. Science, as we know it today, was called "natural philosophy." For the purposes of this book, we will call it science as the more familiar modern term. The word "philosophy" derives from two Greek words: *philia* (love) and *sophia* (wisdom). Literally, it refers to the love of wisdom. The term was adopted by ancient Greek philosophers who wished to emphasize that they did not claim to be wise as much as seekers or lovers of wisdom.

In addition, it is important to appreciate that our modern concept of wisdom is not the same as the ancient concept. Today, wisdom is a relatively narrow concept that we would distinguish from knowledge or skill. However, in the ancient world, wisdom was a more general concept, encompassing all human knowledge, understanding, and skill. In the Old Testament, many references to wisdom (*chakam/chokmah*) refer to the skill of craftsmen:

> Bezalel and Oholiab and every craftsman (*chakam*) in whom the LORD has put skill (*chokmah*) and intelligence to know how to do any work in the construction of the sanctuary shall work in accordance with all that the LORD has commanded. (Exodus 36:1 ESV)

> He was the son of a widow of the tribe of Naphtali, and his father was a man of Tyre, a worker in bronze. And he was full of wisdom (*chokmah*), understanding, and skill for making any work in bronze. He came to King Solomon and did all his work. (1 Kings 7:14 ESV)

God himself is described as using wisdom in his crafting of the world:

> O LORD, how manifold are your works! In wisdom (*chokmah*) have you made them all; the earth is full of your creatures. (Psalm 104:24 ESV)

Solomon was granted wisdom by God, and among other aspects of his wisdom, he speaks of the natural world:

> He spoke of trees, from the cedar that is in Lebanon to the hyssop that grows out of the wall. He spoke also of beasts, and of birds, and of reptiles, and of fish. (1 Kings 4:33 ESV)

Today, we would call Solomon's discussions "biology" and consider him a scientist.

Introduction

When the Bible speaks of wisdom, it refers to the whole scope of human knowledge. It does not refer solely to what we would today think of as wisdom. This is not to say that wisdom in the narrower sense is not valuable. Nor would it be accurate to say that the Bible never uses the word "wisdom" in this narrower sense. Rather, we must keep in mind that when the Bible speaks of wisdom, that concept is often much broader than we might be used to thinking.

With that in mind, consider Paul's words:

> Where is the one who is wise? Where is the scribe? Where is the debater of this age? Has not God made foolish the wisdom of the world? For since, in the wisdom of God, the world did not know God through wisdom, it pleased God through the folly of what we preach to save those who believe. For Jews demand signs and Greeks seek wisdom, but we preach Christ crucified, a stumbling block to Jews and folly to Gentiles, but to those who are called, both Jews and Greeks, Christ the power of God and the wisdom of God. For the foolishness of God is wiser than men, and the weakness of God is stronger than men. (1 Corinthians 1:20–25 ESV)

Paul speaks of the Greeks seeking wisdom. By "Greeks," he is referring to the whole Greek-influenced ancient world. Due to the conquests of Alexander the Great and the dominance of the Roman Empire, Greek culture had spread throughout the ancient Western world. This description of Greeks as seeking wisdom brings to mind the Greeks' own description of themselves as lovers of wisdom, or philosophers.

Remember that wisdom does not indicate our modern narrow concept of wisdom but the whole scope of human knowledge. Indeed, Greek philosophy—Greek wisdom—was a comprehensive worldview incorporating ideas about science, ethics, and theology. It was a worldview that emphasized the seeking of knowledge. Their attitude towards Christianity may be summed up by a pagan philosopher named Celsus, as reported by a Christian apologist, Origen:[1]

> Celsus next proceeds to say, that the system of doctrine, viz., Judaism, upon which Christianity depends, was barbarous in its origin. And with an appearance of fairness, he does not reproach Christianity because of its origin among barbarians, but gives the latter credit for their ability in discovering (such) doctrines. To this, however, he adds the statement, that the Greeks are more skillful than any others in judging, establishing, and reducing to practice the discoveries of barbarous nations

The Greek mind was, at least in its own assessment, open-minded, willing to consider any possibility. They were seeking wisdom and willing to accept it wherever it was found, even among barbarians. Nevertheless, they were the judges who had the skills to decide what was true and what was false.

We can see this attitude playing out in how the philosophers of Athens responded to Paul:

> And they took him and brought him to the Areopagus, saying, "May we know what this new teaching is that you are presenting? For you bring some strange things to our ears. We wish to know therefore what these things mean." Now all the Athenians and the foreigners who lived there would spend their time in nothing except telling or hearing something new. (Acts 17:19–21 ESV)

They were willing to give Paul a hearing. While a few would accept his teaching, most would reject it and scoff at it. The wise of the world found what Paul was saying to be foolishness. This would be the typical response of Greek philosophy to Christianity.

---

[1] Origen, "Contra Celsum," Book 1, Chapter 2.

There are similarities between these ancient Greek philosophical worldviews and the modern scientistic worldview. By a scientistic worldview, I mean a worldview that holds that science is the only way to truth. Those who place their trust in science instead of religion are the intellectual heirs of these Greek philosophers. They display the same confidence in human rationality to discern truth and falsehood. They see their views as deriving solely from reason and experience. They insist that they are open-minded. (In many cases, those who most emphatically claim to have an open mind are in practice the least willing to consider alternatives to their own ideas.) There is nothing new under the sun, and the modern scientistic worldview is actually an ancient idea.

Christianity and Greek philosophy were opposing worldviews. They could not both be true; at most, one could be the correct understanding of the world. Nevertheless, this does not mean that Greek philosophy had no accurate insights. Recall that Greek philosophy was a comprehensive understanding of the world, and included claims about science. But even the insights about science were necessarily interwoven with the false worldview. As such, the Church was faced with the difficult task of discerning the difference between true insights and false worldviews.

The Church and Greek Philosophy

Greek philosophy was not homogeneous; there were a number of different schools of philosophy. These schools disagreed on many subjects, including aspects of natural philosophy or science. A philosophical school would typically combine beliefs about the natural world, ethics, and theology into a complete understanding of the world. The book of Acts records Paul's interaction with a couple of these schools when he was in Athens:

> Some of the Epicurean and Stoic philosophers also conversed with him. And some said, "What does this babbler wish to say?" Others said, "He seems to be a preacher of foreign divinities"—because he was preaching Jesus and the resurrection. (Acts 17:18 ESV)

The Epicureans and Stoics were the two major opposing philosophies at that time. The Epicureans were the precursors to materialists or naturalists and emphasized a belief that everything was made of atoms and void. The Stoics believed instead in a pervading spirit (pneuma) or reason (logos) to the world. Of course, both schools agreed in deeming Paul to be a babbling fool. While these were the major schools, other schools of philosophy also existed, such as Platonism and Pythagoreanism.

The philosophical opposition to Christianity was, early on, divided among these varieties of schools. We can see this in how Basil of Caesarea responds to them:[2]

> The philosophers of Greece have made much ado to explain nature, and not one of their systems has remained firm and unshaken, each being overturned by its successor. It is vain to refute them; they are sufficient in themselves to destroy one another.

Basil sees little point in responding to the philosophers; they could not even agree among themselves. Elsewhere, Basil speaks of the shape of the earth:[3]

> Those who have written about the nature of the universe have discussed at length the shape of the earth. If it be spherical

---

[2] Basil of Caesarea, "Hexaemeron," Homily 1, Section 3.
[3] Basil of Caesarea, Homily 9, Section 1.

or cylindrical, if it resemble a disc and is equally rounded in all parts, or if it has the forth of a winnowing basket and is hollow in the middle; all these conjectures have been suggested by cosmographers, each one upsetting that of his predecessor.

Basil takes there to be a variety of scientific views on the issue of the shape of the earth. From this, and other examples, it is clear that there were a variety of views on scientific issues in the ancient world. It is not as though every student in the ancient world was taught a 2particular scientific model of the world as we do today. Thus, when Christians in this period interacted with the science of their day, they were not interacting with a singular, unified notion but with a diverse collection of beliefs and theories.

However, this would change as the previously dominant schools of philosophy would be replaced by a new school known as Neoplatonism. The word Neoplatonism is a modern attribution; they would have seen themselves as Platonists, following the insights of Plato and Aristotle. Modern scholars term them Neoplatonists because they deem their interpretation of Plato to be somewhat different from what Plato intended. Previously, Platonism had been a minor philosophy, but it became the primary opposition to Christianity in the ancient Western world.

In the realm of science, the views of Neoplatonism were largely inherited from Aristotle. As such, the previous diversity of views was largely replaced by Aristotle's views on science. This meant that Christians were interacting with a singular science of their day instead of a variety of views. They could no longer simply dismiss the philosophers, as Basil did, by pointing out that they did not even agree among themselves.

The political order of the western world would be sent into upheaval by the collapse of the Roman Empire. To those living at the time, it seemed that western civilization itself was collapsing. This caused a reorientation of how the Christian West viewed Greek philosophy. Previously, it had been an opposing worldview. Afterwards, it was the wisdom of a great civilization that had been lost. It became imperative to recapture and learn from its insights. An influential form of this was Scholasticism, which synthesized Christian theology with Greek philosophy in the high Middle Ages.

Gradually, the Christian world began to move beyond the ancient philosophers. Technologies and scientific knowledge began to develop that had not existed in the ancient world. This included the discovery of the Americas, the invention of the telescope, and the development of the printing press. This context helped give rise to the Reformation, which questioned the authority of the Catholic Church and sought to return to the teaching of the Scriptural text. The science of Aristotle was beginning to be questioned but was still largely believed during the first generation of the Reformation. As such, the first generation of reformers became the last generation to interact with Greek philosophy as the science of their day.

We have seen that prior to modern science, the relationship of Christianity with Greek philosophies and science changed over time. Greek philosophy itself transitioned from a diverse set of views into a unified set of ideas. The Church transitioned from viewing the philosophies as the opposition to a source of insight before moving on from them.

Language and the Church

We will be looking at various scientific issues raised by the Biblical text. But which text? Few Christians can read the original languages of the Bible, and instead rely on translations of that text. As such, which translation is being used is often an important issue. In particular, many issues we will look at will depend crucially on the interpretation of key verses. Which translation an author used will change how they understood those verses.

The Old Testament was written mostly in Hebrew. However, by the time of Christ, Hebrew was no longer the common language of the Jews, who instead spoke a related language called Aramaic. The early Church quickly became dominated by Greek-speaking people. The New Testament is written in Greek, and the most prominent very early Church Fathers were all Greek-speaking. This meant they could not read the original Hebrew text of the Old Testament.

Consequently, the early Church relied on a translation of the Old Testament into Greek called the Septuagint. It was translated three centuries before Christ. The name is a reference to a legend according to which it was translated by seventy (or seventy-two) translators who miraculously all came up with exactly the same translation independently. It originally only consisted of a translation of the Pentateuch, or the first five books of the Bible. Over time, additional translations were added of other books in the Hebrew Old Testament, as well as other books not accepted as canonical by the Jews. These additional books are now known as the Apocrypha.

The Church also grew in geographic regions under heavier Roman influence, where the common language was Latin rather than Greek. These Christians were at a disadvantage because they could read neither Greek nor Hebrew. A variety of translations of varying quality were made from the Septuagint and the books of the Greek New Testament into Latin. Today, these are known collectively as the *Vetus Latina*, which means Old Latin.

However, scholars determined that the Septuagint did not always agree with the Hebrew text. It contained blatant mistranslations and differences. Further, the *Vetus Latina* translations were often of poor quality. This would lead the early Church Father, Jerome, to produce his own translation of the New and Old Testaments into Latin, producing what we now know as the Vulgate. He included the Apocrypha, applying the label "Apocrypha" to those texts to separate them from the canonical texts. While Jerome did translate the Old Testament from the original Hebrew, his translation was often influenced by the Septuagint.

The Greek/Latin language divide would contribute to the development of different churches with different theologies that would eventually break communion with each other. Thus, Christianity would be divided along linguistic lines into the Eastern Greek-speaking Church and the Western Latin-speaking Church. Much of the Greek-speaking world would be conquered by Muslim invasions, greatly diminishing the Eastern Church. Today, the vast majority of Christians belong to the Western Church, either through the Roman Catholic Church or the Protestant churches that broke away during the Reformation.

The Greek-speaking Church adopted the Septuagint as the official text of Scripture. The Latin-speaking Church adopted Jerome's Vulgate as the official text of Scripture. The Reformers insisted on going back to the Hebrew text instead of a translation. This may seem obvious, but it should be kept in mind that the Hebrew text we have today, known as the Masoretic Text, was standardized into its current form during the Middle

Ages. The Septuagint and Vulgate, while both translations, are much older than the Masoretic text and based on much earlier manuscripts. Thus, some argue that the Septuagint is more reliable than the Masoretic text. Nevertheless, even the Roman Catholic Church translates from the Hebrew in modern translations.

It is crucial to understand the weight of the Septuagint. It was widely believed to have been miraculously translated and was thus viewed as an inspired text. The early Church had used it since the days of the apostles. Furthermore, most Christians would have been unable to read the Hebrew text even if they had wanted to. For all intents and purposes, the Septuagint was Scripture. In fact, when discrepancies were found between the Hebrew text and the Septuagint, early Christians claimed that the Hebrew text had been corrupted.

As we consider some of the specific issues discussed historically, we will find a number of cases where the Septuagint's translation causes scientific difficulties. In some cases, it appears that the translators of the Septuagint allowed the ideas of Greek philosophy to affect their translations. Historical Christians rarely resolved these difficulties by appealing to the Hebrew text. This is partially due to ignorance; they did not know what the Hebrew text said. However, it was also due to the high esteem in which they held the Septuagint.

Questions

As we look through these issues in the upcoming chapters, there are three questions that should be kept in mind. Firstly, should we be encouraged or discouraged by the Church's historical handling of scientific issues? We should, in fact, be encouraged. We're not the first to face scientific questions about the text; similar questions have been a theme throughout the history of the Church. Furthermore, the Church has done well in its handling of these scientific issues. It has rejected the science of the day when it was incompatible with the Biblical text and, for the most part, has avoided unnecessary conflict.

Secondly, how did historical Christians understand the authority of the Biblical text, especially in areas of science? We find that the authority of the Biblical text was unquestioned. Historical Christians thought they could draw conclusions about scientific issues from the Biblical text. They always insisted that the Biblical text was correct, even if sometimes it was misinterpreted.

Thirdly, what was the underlying cause of mistakes and errors? We find that problems usually stemmed from not going back to the source or sources. Issues arose when using flawed translations of the text, reading flawed scientific theories into the text, engaging in dubious exegesis, or failing to re-evaluate the evidence for a scientific theory or claim. Problems are resolved by going back to the text to carefully evaluate what the original authors of the Biblical text intended as well as going back to the scientific data to carefully evaluate whether it supports a particular claim. If you do not go back the source, you will quickly find yourself defending a flawed exegesis or flawed scientific theory.

These three questions and their answers reveal the key lessons that can be learned from a study of the historical interaction of science and faith.

Chapter 2

Are the Heavens Immutable?

The Scientific Question

Are the heavens—what we moderns call "outer space"—made of the same "stuff" as the earth? Today, we take it for granted that the rest of the universe is made up of atoms and molecules of a nature identical to those found on earth. The laws of physics apply equally well in other galaxies as they do here on

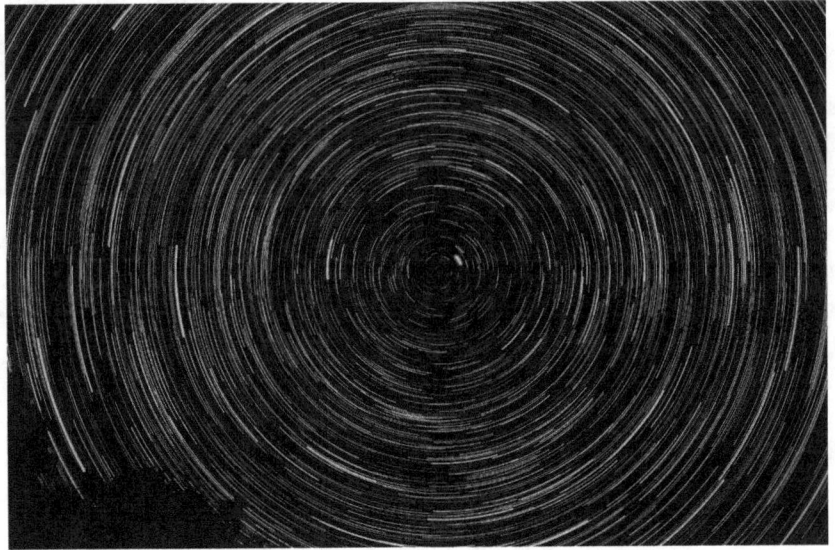

Figure 2.1: Time lapse image of the stars

earth. Fundamentally, the whole physical universe is made of the same stuff and follows the same rules.

However, the classical world did not take this for granted. They widely agreed that everything on earth was made up of four elements: earth, water, air, and fire, and that the properties of any earthly thing could be explained by the combination of elements that comprised it. However, Aristotle proposed that the heavens were not made of these four elements but of a fifth element known as *aithēr*, or aether. Aether is not to be confused with the ethers, a class of chemical compounds. Aether had unique properties, which explained the unique nature of the heavens. In particular, aether had the property of immutability; it did not change.

The idea of postulating that the heavens were made of a distinct element seems very strange to us. However, it is not so different from when modern cosmologists postulate the existence of dark energy or dark matter. These modern theories postulate entities to account for differences between what we would expect from our understanding of physics and our observations of distant space. Likewise, there were a number of ways in which the movement of the heavens differed from observed movements on earth, leading to the postulation of a unique element.

A dropped rock will fall to the ground, but stars do not fall from the heavens. According to classical science, this is because a rock is mostly made up of heavy elements. Smoke, on the other hand, rises because it is mostly made of light elements. But stars do neither of these things. They do not fall to the ground like rocks or float away like smoke. They remain an unchanging distance from the earth.

However, the stars do move. In the course of a night, they will all complete a rotation around the north pole. But the movement is circular. Unlike linear movement, circular movement is not necessarily finite. Since a circle feeds into itself, circular movement can continue indefinitely. Thus, stars move in a way that could, in principle, continue forever.

Furthermore, this circular movement keeps a constant pace. The stars are neither speeding up nor slowing down. If the stars were slowing down, it would seem they would eventually stop. If they were speeding up, there must eventually be some limit to that speed. Instead, the stars move at an unchanging speed.

Additionally, there are never any new stars. Nor do any stars disappear. Some stars are visible at different times of the year, but the same stars always return. In fact, the same stars have been in the sky as long as people have been watching them. The population of the stars remains unchanging.

The unifying feature of the heavens was their lack of change. The stars did not either fall to the earth or rise. They moved in an unchanging circular motion at an unchanging speed. The population of the stars remained constant, never changing. While things on earth were always in flux and subject to change, the heavens were the realm of immutability.

In the classical conception, most objects were made up of some combination of the four elements. A rock, for example, would primarily consist of earth, a heavy element. However, it would also contain some of the other elements. But these different elements had different tendencies in terms of how they moved. If the heavens were made up of a combination of elements with such different tendencies, they could not be stable and would eventually break apart.

These considerations are what prompted Aristotle to postulate that the heavens were not made of the ordinary elements found on earth but instead of their own unique element, aether. Aether, rather than naturally falling towards the center of the universe (the earth), naturally moved in a circle around it. Furthermore, it could not be created, destroyed, damaged, or changed in any way. It was immutable. This unique element explained why the heavens were so different from the earth.

This theory had an important implication. If aether cannot be created or destroyed, this implies that the heavens are eternal. According to Aristotle, the heavens had existed into an eternal past and would exist into an eternal future. They were not brought into existence either by natural forces or an act of divine fiat. Neither would they grow old or eventually cease to exist.

The Biblical Discussion

There are three major issues of potential conflict between the Bible and the heavens being made of immutable aether: the creation, destruction, and decay of the heavens. Aether was postulated to be uncreateable, indestructible, and not subject to decay. On the other hand, the Bible appears to speak of the creation, destruction, and decay of the heavens.

The first issue is the creation of the heavens at some point in the finite past. According to Aristotle, the heavens were eternal and had not been created; they had simply always existed. However, the Bible frequently speaks of God as the creator of the heavens and the earth.

> You are the LORD, you alone. You have made heaven, the heaven of heavens, with all their host, the earth and all that is on it, the seas and all that is in them; and you preserve all of them; and the host of heaven worships you. (Nehemiah 9:6 ESV)

> My help comes from the LORD, who made heaven and earth. (Psalm 121:2 ESV)

> And when they heard it, they lifted their voices together to God and said, "Sovereign Lord, who made the heaven and the earth and the sea and everything in them, (Acts 4:24 ESV)

It is conceivable that God could have chosen to create an eternal heavens; however, the first verse of the Bible states:

> In the beginning, God created the heavens and the earth. (Genesis 1:1 ESV)

God is not simply the creator of the heavens, he created them at a particular point in time, the beginning. Additionally, Proverbs describes wisdom as existing before the heavens:

> When he established the heavens, I was there; when he drew a circle on the face of the deep, (Proverbs 8:27 ESV)

If the heavens existed eternally, there could not be wisdom before the heavens because the heavens had always been there. It is very difficult to see how one could deny that the Bible teaches a historical creation of the heavens.

However, one could take the position that the heavens are made of an immutable fifth element that God created. God is omnipotent and thus able to create what would otherwise be uncreatable. As an analogy, while modern scientists believe that energy cannot be created or destroyed, nobody sees this as in conflict with believing that God is the ultimate creator of all energy.

The second issue is the destruction of the heavens. Aristotle held that the heavens had not only always existed just as they are now; so too they would also continue to exist into an eternal future. However, there are a number of passages that speak of the destruction of the heavens as part of a future judgment:

All the host of heaven shall rot away, and the skies roll up like a scroll. All their host shall fall, as leaves fall from the vine, like leaves falling from the fig tree. (Isaiah 34:4 ESV)

Truly, I say to you, this generation will not pass away until all these things take place. Heaven and earth will pass away, but my words will not pass away. (Matthew 24:34–35 ESV)

But the day of the Lord will come like a thief, and then the heavens will pass away with a roar, and the heavenly bodies will be burned up and dissolved, and the earth and the works that are done on it will be exposed. (2 Peter 3:10 ESV)

However, there has been a long-standing disagreement within the church about how to understand this future judgment. Some believe that the Scriptures teach a future total annihilation of the physical universe. Others believe that the universe will undergo some sort of transformation, but will not be utterly destroyed. Still others believe that these references to the heavens and the earth are intended symbolically and do not have the physical universe in view. It would not serve our purpose to digress from our subject into the contentious field of eschatology. We must simply acknowledge that there is wide disagreement on this issue, both today and throughout church history.

If the heavens are to be destroyed or even just transformed, this would appear to conflict with the existence of immutable aether. However, just as we could argue that God could create immutable ether, so too could he destroy it. God is omnipotent and thus not constrained by the ordinarily indestructible nature of aether.

The third issue is the decay of the heavens—the idea that the heavens are growing old. This contradicts the claim that the heavens are immutable and not subject to change. Paul writes that all of creation is under bondage to corruption:

> For the creation was subjected to futility, not willingly, but because of him who subjected it, in hope that the creation itself will be set free from its bondage to corruption and obtain the freedom of the glory of the children of God. For we know that the whole creation has been groaning together in the pains of childbirth until now. (Romans 8:20–22 ESV)

Other passages speak of both the heavens and the earth undergoing a process of growing old and decaying.

> Of old you laid the foundation of the earth, and the heavens are the work of your hands. They will perish, but you will remain; they will all wear out like a garment. You will change them like a robe, and they will pass away, (Psalm 102:25–26 ESV)

> Lift up your eyes to the heavens, and look at the earth beneath; for the heavens vanish like smoke, the earth will wear out like a garment, and they who dwell in it will die in like manner; but my salvation will be forever, and my righteousness will never be dismayed. "Listen to me, you who know righteousness, the people in whose heart is my law; fear not the reproach of man, nor be dismayed at their revilings. For the moth will eat them up like a garment, and the worm will eat them like wool, but my righteousness will be forever, and my salvation to all generations." (Isaiah 51:6–8 ESV)

Additionally, Hebrews 1:10–12 quotes Psalm 102:25–27.

These passages are in direct conflict with the Aristotelian conception of the heavens as perfect and unchanging. The Biblical texts extend the concept of the corruption of the world to all of creation, not just the earth. It might be argued that the passages are metaphorical. However, it would be odd to use the decaying heavens as a metaphor for something else if the heavens are not, in fact, decaying.

Another possibility would be to argue that the word "heavens" in these passages does not refer to what we would call outer space but to the atmosphere. The Bible does refer to the atmosphere as the heavens, although this is not as clear in our English translations. For example, in many places the Biblical text refers to the "birds of the heavens," which most translations render as "birds of the air." In Biblical usage, the heavens are typically everything above us, including both the atmosphere and outer space. However, the phrase "heavens and earth" is frequently used to refer to everything God created. It would be odd for these texts to use this phrase to refer only to the earth and the lowest part of the heavens.

This issue cannot be avoided by distinguishing between eternality and immutability. An omnipotent God could create or destroy the immutable heavens. As such, it is possible to reconcile immutable heavens with non-eternal heavens. Nevertheless, an omnipotent God does not help reconcile a belief in immutable heavens with statements that they grow old. If the text teaches that the heavens are growing old, this is in unavoidable conflict with the claim that the heavens do not change by natural processes in any way.

Each of these issues poses a challenge to reconciling Biblical teaching with aether. The creation, destruction, and decay of the heavens all give reasons to see the Bible as being in conflict with this scientific theory. However, if one is willing to separate the immutability of the heavens from their eternality and understand many references to the heavens as metaphorical or limited, it would be possible to resolve the conflict. Nevertheless, it must be acknowledged that this resolution is awkward, and the text is much more readily understood as being in conflict with aether.

Historical Christians

Origen of Alexandria

Origen referenced this subject in his apologetic against the philosopher Celsus. The context is that Origen has just argued that, based on Celsus's own argument, it must be concluded that human bodies are the same as the heavenly bodies because they are made of the same material. However, he anticipates that Celsus could avoid this implication by invoking Aristotle's theory of an immutable heavens.[1] Origen points out that both the Platonists and the Stoics "nobly protested" this idea. Furthermore, Origen states that Christians also cannot accept aether on the basis of a quotation from Isaiah 34, which speaks of the heavens growing old and being changed. Clearly, Origen sees aether as inconsistent with Biblical teaching.

Basil of Caesarea

Basil speaks of the essence of the heavens, that is, the material from which the heavens are made, as being like smoke.[2] This comes from Isaiah:

> Lift up your eyes to the heavens, and look at the earth beneath; for the heavens vanish like smoke, the earth will wear out like a garment, and they who dwell in it will die in like manner; but my salvation will be forever, and my righteousness will never be dismayed. (Isaiah 51:6 ESV)

However, where our English translation indicates the heavens in the future will vanish like smoke, Basil takes it as a description of the substance from which the heavens are formed. The reason for Basil's reading can be ascertained by reading the Septuagint's translation of this verse:

---

[1] Origen, "Contra Celsum," Book IV, Chapter 56.
[2] Basil of Caesarea, "Hexaemeron," Homily I, Section 8.

Lift up your eyes to heaven and look at the earth beneath, because heaven was strengthened like smoke, and the earth will die like these things, but my salvation will be forever, and my righteousness will not fail (Isaiah 51:6 New English Translation of the Septuagint)

The Hebrew text "the heavens vanish like smoke" became "heaven was strengthened like smoke." In Hebrew, the word "vanish" is "*malach*" and is only used a handful of times. Elsewhere, it refers to adding salt to something. Our modern English translations universally understand this as the heavens dissolving like salt. However, the Septuagint translator appears to have had difficulty with this word and rendered it using στερεόω (stereoō), which means to make solid or to strengthen. This was probably influenced by the translation of the firmament, or expanse of the heavens, from the Genesis 1 creation account as στερέωμα (stereōma). Nevertheless, it makes little sense to say that the heavens were strengthened like smoke because smoke is not generally considered strong. Basil reads the verse more like "strengthened from smoke," arguing that God created the heavens from a smoke-like substance. He states that this substance is without solidity or density.

Later in his homily, Basil returns to a discussion of the nature of the heavens.[3] Basil describes the scientific theorizing of his time as a "noise of words." He first mentions that some say that the heavens are made up of ordinary elements. Then he mentions the theory of the fifth element, aether. He describes aether as an "element after their own fashioning."

Basil explains two reasons motivating the postulation of aether. One is the heavens' circular rather than linear movement. The other is the tendency the heavens would have to

---

[3] Basil of Caesarea, Homily I, Section 11.

26

dissolve if they were made of disparate elements with different natural movements. Basil dismisses the whole discussion as frivolous. As he describes it, each philosopher overturns the system of the last, leaving nothing certain and rendering the whole exercise fruitless. He insists that instead we should be content with what Moses said: that God created the heavens and the earth. Even the brightest mind cannot understand the least natural phenomenon or give a suitable explanation of it.

Basil does not explicitly reject aether, either Scripturally or scientifically. He does use dismissive wording such as "noise of words," "element of their own fashioning," and "frivolities." Nevertheless, he does not explicitly reject it. His point is the discord between different philosophers, not the merits of any individual theory. His claim that the heavens were made of a smoke-like substance would seem to be in conflict with either theory of the heavens, but Basil does not draw our attention to this conflict.

Ambrose of Milan

Ambrose also discusses this question in his work on creation.[4] He mentions there being two major theories about the heavens: either the heavens are made of the same four elements found on earth, or a unique fifth element. He explains the reason for postulating aether as being due to the tendency of any composite of elements to eventually lead to the destruction of the heavens. Aether is postulated to explain how the heavens can remain without such destruction being apparent.

Ambrose declares that this opinion "could not withstand the words of the prophet" and quotes Scripture to show that the heavens and earth are not immutable but will in fact pass away. In particular, he quotes Psalm 102 and Matthew 24. Since

---

[4] Ambrose of Milan, "Hexameron," Book 1, Chapter 6, Section 23-24.

the heavens will pass away, they are clearly not made from this immutable material.

Ambrose goes further to argue that aether will not solve the problem. Aether is very different from the other elements and adding a dissimilar part to a system will usually make it less stable. Thus, it would seem that aether would not help the heavens last since it is completely dissimilar to the other elements in the universe.

John Chrysostom

Chrysostom comments on Paul's statement that "the creature was made subject to vanity" from Romans 8:20.[5] He interprets this as Paul referring to the whole creation becoming "corruptible." This corruptibility is because of man's sin. The curse applies not only to man but also to the earth and heavens. The heavens will grow old and be changed. He references both Psalm 120 and Isaiah 40 to support this contention.

Chrysostom does not refer to aether. Indeed, he shows no awareness of the theory. Nevertheless, he takes the explicit position that the heavens grow old and will be transformed in the future. Thus, he takes a position that would conflict with the theory of aether.

John Philoponus

John Philoponus spent much of his work, Against *Aristotle on the Eternity of the World*,[6] arguing against the immutable heavens. He sought to show that the heavens were made of the same material or elements as were found on earth. Sadly, much of the book has been lost, and is now only available in various fragments.

---

[5] Chrysostom, "Homilies on Romans," Homily 14.
[6] Philoponus, *Against Aristotle on the Eternity of the World*.

In a simple geocentric model, the earth is at the center of the universe and all of the heavens are moving in a circle around that center. However, when we begin to look closer, not everything lines up with this simple model. While everything in the heavens rotates, a few stars typically rotate a little slower than the rest of the stars. Once the rest of the sky rotates back to its starting position, these stars will be in a different position relative to other stars. If you watch their movement from one night to the next, they trace their own circle around the night sky. However, sometimes these stars will slow down and reverse course. They will travel in the opposite direction along their circle for a time before resuming their regular course. This was called retrograde motion.

The Greeks called these these special stars *planētai* (meaning "wanderers"), from which we derive our name for them, "planets." Today, our conception of a planet is an earth-like body orbiting the sun or another star. However, to the naked eye, they look much like stars, only distinguished by their unusual movements. The circle that the planets trace around the sky is the orbit they are making around the sun. The retrograde motion or apparent reversal of direction of a planet is caused by the earth passing the other planet in its orbit, the way that a car in a neighboring lane may appear to be moving backwards as you pass it.

A more complex geocentric model was required in order to explain these strange movements. This involved rotating spheres that were not centered on the earth, termed "eccentric spheres." It also involved epicycles, a system where the planets did not orbit around the earth directly, but around a moving point that itself orbited around the earth. We will not go into detail about how these worked, but they were highly accurate for their time in predicting the positions of the planets.

Philoponus uses this state of affairs to argue against the theory of aether[7]. He points out that the stars do not follow a simple circular motion around the center of the universe. As such, it does not make sense to claim that the heavens are made of a material that has such a simple circular motion. Indeed, Aristotle's system was compelling because it explained the heavens in a simple fashion. However, this is undermined by the additional complexity required in order to explain planetary movements.

Philoponus ends with the observation that the stars move up and down. They go from perigee, when they are closest to the earth, to apogee, when they are farthest away. This means that these stars have the up (away from the earth) and down (towards the earth) movements that are supposed to be properties of the terrestrial elements and not aether. The fact that these stars, actually the planets, vary in how far away they are from the earth can be seen in that they vary in how large and bright they appear. But this further shows that the stars do not move in simple circles.

Aristotle's argument postulated that the heavens were neither heavy nor light. This is why they neither fell down to the earth nor flew away from it. If they were made of ordinary elements, why did they not do either of these? Philoponus answers this objection.[8] He considers two possibilities: either the heavens are light and thus should be rising away from the earth, or they are heavy and should be falling down to the earth. If the heavens are light, he argues that the universe is finite and thus there is nowhere higher up for the heavens to rise into, comparing this to the idea that the earth is at the

---

[7] Philoponus, Book 1, Fragment 7.
[8] Philoponus, Book II, Fragment 47.

bottom of the universe. Just as there is nowhere farther down than the earth, there is nowhere farther up than the heavens.

However, if they are heavy, they cannot move down because there is no room for them. That space is already occupied. Even if there was empty space, the heavens would not fit into it because they are very large. Furthermore, Philoponus argues that the heavens consist of a rigid form that is not easily broken. Thus, they cannot break apart to fit into the smaller space below.

In another passage, Philoponus addresses the argument that the heavens have never been observed to change.[9] Philoponus points out that many things on earth can also last a long time. Some animals live a long time. Diamonds and mountains last a very long time. In particular, he mentions Mount Olympus, the highest mountain in Greece. According to Greek mythology, it was where the gods lived. Mount Olympus, like the stars, was not seen to have changed in size over time.

Philoponus draws an analogy between an animal and the world. As long as they exist, their most important parts must remain intact. The heavens are, according to Philoponus, the most important part of the world. He believes that their motion is responsible for guiding the natural movements here below. Thus, they must be preserved as long as the world exists.

These are just a sampling of the arguments that Philoponus put forward. As the Stanford Encyclopedia of Philosophy describes:[10]

Philoponus succeeds in pointing to numerous contradictions, inconsistencies, fallacies and improbable assumptions in Aristotle's philosophy of nature relating to these claims. Dissecting Aristotle's texts in an unprecedented way, he time

---

[9] Philoponus, Book IV, Fragment 80.
[10] Wildberg, "John Philoponus."

and again turns the tables on Aristotle and so paves the way for demonstrative arguments for non-eternity.

Philoponus delivered a withering critique of the theory of aether.

Thomas Aquinas

Aquinas considers the claim that Aristotle's theory proves that the world is eternal.[11] The second objection he considers states that there are incorruptible things in the world, including the heavenly bodies. Something that is incorruptible or unchanging can neither come into existence nor pass out of existence. Since the heavens were thought to be unchanging, they must have existed eternally.

Aquinas accepts the assumption that the heavenly bodies are incorruptible. However, he rejects the argument that this implies their eternality. He argues that the heavens were given the ability to exist forever when they were created. Hence, they did not exist prior to that point. Their ability to exist forever only extends after that point, not before it. He agrees that the heavens could not have come into existence in the ordinary way or natural mode.

In the third objection, Aquinas considers the argument made by Aristotle that heaven is ungenerated because "it has no contrary from which to be generated." The argument was that while upward movement was contrary or opposite to downward movement, there was no opposite or contrary to circular movement. In order for the aether to be destructible in the classical understanding, it would require another element that moved in the opposite direction; however, Aristotle

---

[11] Aquinas, *The Summa Theologiæ of St. Thomas Aquinas*, First Part, Question 46, Article 1.

argued that there could be no opposite to circular movement. This implied that the aether was indestructible.

Aquinas accepts the logic of Aristotle concerning the nature of the heavens and his conclusion that the heavens are immutable. However, he argues that this only applies to the normal processes of change. In creation, God could create what could not come into existence naturally.

Aquinas handles both of these objections in a similar fashion. He accepts the immutability of the heavens while rejecting their eternality. He insists that they were created and that Aristotle's logic does not show their eternality.

Later, he discusses whether or not the fire described in 2 Peter 3 will destroy the heavens.[12] Aquinas cites a statement that the heavens cannot be affected by outside forces, implying that they are incorruptible. If the heavens are incorruptible, nothing can affect them. They certainly cannot be burned or melted.

He also cites a gloss, which are notes written in some copies of the Biblical text akin to what would be found in a modern study Bible. According to this gloss, the fire will only reach as high as the flood, some 15 cubits higher than the mountains.

Next, Aquinas seeks to explain why it makes sense for the terrestrial elements to be destroyed by fire while the heavens remain intact. He argues that the cleansing fire will bring all things to perfection; however, what is required for this is different for different bodies. The terrestrial elements have been mixed together and are thus no longer pure. They will have to be thoroughly burned. However, the heavenly elements are not mixed. Since they are immutable, they have not been damaged in any way. There is nothing wrong with the aether in heaven that would need to be cleansed. However, the heavens are

---

[12] Aquinas, Supplement, Question 74, Article 4.

made perfect by the cessation of their movement. Aquinas goes into more detail on this elsewhere, but he believes that while the heavens will remain intact, they will cease to move.

In the first objection, Aquinas considers Psalm 102:25–26, which was discussed in the introduction to this chapter. It speaks of the heavens perishing, which would seem to imply that they will be destroyed. Aquinas appeals to Augustine[13] who argued that they refer to the lower heavens or atmosphere rather than the whole universe. Augustine's reason for this comes largely from his interpretation of 2 Peter 3, which will be considered in the next objection. Alternatively, Aquinas argues that it might refer to a cessation of movement rather than an outright destruction of the heavens.

The second objection argues that 2 Peter 3 indicates a fire will consume the heavens. Aquinas answers by invoking an earlier passage in that chapter.

> For they deliberately overlook this fact, that the heavens existed long ago, and the earth was formed out of water and through water by the word of God, and that by means of these the world that then existed was deluged with water and perished. But by the same word the heavens and earth that now exist are stored up for fire, being kept until the day of judgment and destruction of the ungodly. (2 Peter 3:5–7 ESV)

Peter speaks of the world perishing in that flood. However, the whole universe did not perish, but only the area up to the height of the flood. Consequently, Aquinas concludes that the heavens referred to by Peter only include the area under the floodwaters. This is also the reasoning of Augustine. Indeed, the idea that the future fire only goes as high as the floodwaters appears to have its origin in this passage.

---

[13] Augustine, "The City of God," Book XX, Chapters 18 and 24.

Aquinas's reading does seem awkward. On the other hand, it must be acknowledged that Aquinas is not simply inventing an odd interpretation. Previous authors had given the same interpretation. Furthermore, his reading is rooted in the text.

Aquinas accepts aether but attempts to reconcile it with Christian belief. On the creation of the heavens, he seeks to argue that the immutability of the heavens is not incompatible with their creation. God can create outside of ordinary natural processes. On the destruction of the heavens, he limits the scope of that destruction. He thinks the destruction of the world only goes as far as the flood or that it refers only to the cessation of movement. He does not discuss the question of the decay of the heavens.

Martin Luther

Luther briefly mentions the issue of the fifth element.[14] He simply notes that some philosophers held to the existence of a fifth element. However, he states that all philosophers agree that the heavenly realm, consisting of eight spheres corresponding to the sun, the moon, the five known planets, and the fixed stars, is incorruptible. While the existence of aether was uncertain, Luther appears to accept without question that the heavens are incorruptible and unchangeable.

John Calvin

Calvin rejects the immutability and incorruptibility of the heavens.[15] He quotes Paul in Romans 8:22 describing the whole creation as subject to corruption. Thus, Calvin argues that the heavens are indeed subject to corruption on account of the fall of man, rejecting the idea of an immutable heavens. However,

---

[14] Luther, *Luther On The Creation: A Critical and Devotional Commentry on Genesis*, 1:Part II.

[15] Calvin, *Calvin's Commentary on the Bible*, Commentary on Psalm 102:25.

he notes that others consider it absurd to claim that the heavens are subject to corruption and thus not immutable. They interpreted the statement conditionally, as if it said not "the heavens shall perish" but instead "the heavens shall perish if it so please God." Thus, it does not say the heavens will be destroyed but only that God could destroy the heavens if he chose to. Calvin rejects the conditional understanding of the claim, saying that such a reading rather obscures what the text is saying.

Conclusions

The Christian Interaction

Most of the authors we have looked at rejected the immutable heavens. Origen and Ambrose both explicitly identify Aristotle's theory of aether and reject it as contrary to Scripture. Calvin does not explicitly identify aether but clearly rejects the claim that the heavens are immutable despite the fact that some thought such a position absurd. Philoponus does not discuss the Scriptural part of the question, but goes to great effort to defend the mutability of the heavens scientifically. Chrysostom teaches that the heavens are mutable but shows no awareness of scientific claims to the contrary. Basil is dismissive of any theory of the heavens but does not explicitly reject any particular theory, including aether.

On the other hand, some authors did not reject the immutable heavens. Luther references aether as in question, but appears to accept the immutability of the heavens without much consideration. Aquinas, on the other hand, argues for the compatibility of the immutable heavens with Christian belief. It should be noted that he does not actually argue that the heavens are immutable, but only answers objections based on Aristotle's theory without challenging that theory. He rather sought to show that the theory was compatible with Scripture but did not explicitly endorse it.

Most of the Biblical discussion concerns the fate of the heavens, whether they will be destroyed or grow old. Indeed, Aquinas is the only author who considers whether the immutability of the heavens precludes their creation. The authors who reject aether as unscriptural do so on the basis of the Scripture teaching that the heavens will grow old or be destroyed. It is worth noting that they display no qualms about applying the words of a prophetic passage to settle a scientific question.

All of these authors believed in the creation of the heavens. Indeed, we will see later that they held to creation *ex nihilo*, which implies the creation of the heavens. Even Aquinas, who otherwise accepts Aristotle's cosmology, rejects him on the question of eternality. He insists, with all the church, that the heavens were created by God in history.

Modern Science

In 1572, a new star appeared in the sky, remaining there until it faded away in 1574. In 1604, another new star appeared, lasting for almost a year before also fading away. This was, to the surprise of European observers, a change in the heavens. However, there was a ready Aristotelian response: these new stars were not actually stars, but some phenomenon in the atmosphere. Nevertheless, the new stars moved like stars, along with the general motion of the heavens, and not like the nearer objects such as the sun, moon, or planets. Indeed, any object closer than the stars should have moved relative to those stars, but the new star did not. Tycho Brahe, who took a special interest in the first new star, argued that the new star had always been there but had previously been hidden by God. Today, we understand that these new stars were stellar explosions, which we call supernovas. The word "nova" means "new" and is a reference to the original belief that these were new stars. Supernovae had not been noticed before because they are extremely rare. There have been only eight supernovae visible to the naked eye recorded in human history.

During the same time, the telescope came to prominence as a tool for observing the heavens. This allowed details to be observed about the heavens that had never been seen before. Some of these details demonstrated imperfections in the supposedly perfect, immutable heavens. Notably, Galileo Galilei observed craters on the moon and spots on the sun.

These observations led to the abandonment of the idea of the immutable heavens. It is now essentially universally accepted that the heavens are made of the same basic stuff as the earth. There are more people who believe in a flat earth than in an immutable heavens. Aether, as conceived by Aristotle, does not exist. The term aether has been reused in more modern theories, but in reference to an entirely different entity. Furthermore, none of these theories are currently widely accepted.

Evaluation of the Response

Aristotle's theory turned out to be incorrect. The heavens are made of the same stuff as the earth. They are not immutable and are subject to decay. This means that the Christians who rejected the immutability of the heavens were correct. They would turn out to be on the right side of history, even if, for many of them, it took a millennium for that history to happen. Aquinas, on the other hand, would turn out to be reconciling Christian belief with an incorrect scientific theory.

However, Aquinas did not argue that the Scriptures taught an immutable heavens. He merely argued that the immutable heavens were compatible with the Scripture's teaching. The fact that the heavens turned out to not be immutable does not mean he was incorrect about the incompatibility. It is conceivable that the Scriptures might be consistent with either a mutable or immutable heaven.

Back towards the beginning of this chapter, we discussed the Scriptural passages that relate to this question. If one is willing to separate the eternality of the heavens from their immutability and to interpret some references to the heavens as metaphorical or limited, it is possible to read these passages in a way that is consistent with aether. However, such a reading is awkward and difficult to defend as the original author's intention. Certainly, if aether were the modern scientific consensus, these passages would be urged as demonstrations of scientific error in the Biblical text.

We see that it is not a new phenomenon for there to be conflict between Scripture and science on questions regarding the origin of the universe. The Church insisted, on the basis of the authority of Scripture, that the heavens were not eternal. Most historical Christians further insisted that they were not immutable but instead subject to decay. The correct answer turned out not to be reconciling the text with the theory but going back to the scientific data and reevaluating it.

Chapter 3

Is Matter Conserved?

The Scientific Question

Consider the construction of a house. It consists of taking materials such as wood, metal, and concrete and imposing the form of a house on those materials. The materials continue to exist, but they have lost whatever form they used to have and now have the form of a house. Those materials will be there as long as the house persists. Once the house is demolished, the materials will lose the form of a house but will persist in some other form. Building a house does not create the raw materials, nor does the demolition of the house destroy them.

What if, instead of being demolished, the house was burned down? At first glance, it would seem that this would result in the destruction of the materials. However, this is not exactly the case. The materials are not destroyed but instead, as the ancients saw it, transformed into fire, smoke, and ash. In fact, every process in the natural world can be seen to involve a transformation of materials from one form into another, not the creation or destruction of those materials. Classical science expressed this idea with the Latin phrase *"ex nihilo nihil fit"* which means nothing comes from nothing.

Classical science held that everything below the heavens (specifically, the moon) was made out of four elements: earth, water, air, and fire. But it was possible to transform one element into another. For example, Aristotle thought that when water evaporated, it turned into air. When it rained, the air turned back into water.

But how can air turn into water unless air and water are made up of the same underlying material? This gives rise to the idea of "prime matter," the underlying material from which all things are made. This matter may change from one element to another. It may lose or take on various forms. But the matter itself is not created or destroyed. In modern terminology, we would call this the conservation of matter.

But this leads to a question: if matter cannot be created or destroyed, where did it come from? If we take the conservation of matter as an absolute, we must conclude that matter was never created. Instead, it must exist into an infinite past. Furthermore, since matter cannot be destroyed, it must also exist into an infinite future. If the conservation of matter is absolute, then matter is eternal.

This is related to the issue discussed in the previous chapter as to whether the heavens were eternal. Clearly, if prime matter has only existed for a finite amount of time, then the heavens must also be finite and not eternal. However, the converse is not true. Conceivably, the heavens could have been made out of pre-existing eternal prime matter. If so, the heavens themselves would not be eternal, but the prime matter would be.

The Biblical Discussion

The Christian doctrine of creation *ex nihilo* teaches that God created the world out of nothing. He did not create the world *ex materia*—out of any sort of pre-existing eternal matter. This is a doctrine, like the trinity, that has been universally accepted by orthodox Christianity but for which no simple and unequivocal proof text can be provided. This is not to say that the Bible does not teach the trinity or creation *ex nihilo*, but that the Bible does not explicitly articulate those doctrines in the same way as an orthodox Christian would.

The Bible actually teaches that God created the world from pre-existing matter. Prior to God declaring, "Let there be light," there was already a formless earth and waters:

> The earth was without form and void, and darkness was over the face of the deep. And the Spirit of God was hovering over the face of the waters. (Genesis 1:2 ESV)

God formed animals, not out of nothing, but out of the ground:

> Now out of the ground, the LORD God had formed every beast of the field and every bird of the heavens and brought them to the man to see what he would call them. And whatever the man called every living creature, that was its name. (Genesis 2:19 ESV)

Peter possibly refers to the whole world being created out of water:

> For they deliberately overlook this fact, that the heavens existed long ago, and the earth was formed out of water and through water by the word of God, (2 Peter 3:5 ESV)

Thus, the Bible clearly teaches the creation of the world from previous material. Nevertheless, this is not incompatible with the doctrine of creation *ex nihilo*. Rather, that doctrine teaches that God created the primordial waters or matter prior to making all things from them. God created the world from pre-existing matter, but only after first creating that matter.

43

But does the Genesis creation account teach that God first created the formless and void matter? The dispute on this question depends on the correct translation of the first few verses of the Bible.

According to many English translations, it says something like:

> In the beginning, God created the heavens and the earth. The earth was without form and void, and darkness was over the face of the deep. And the Spirit of God was hovering over the face of the waters. And God said, "Let there be light," and there was light. (Genesis 1:1–3 ESV)

However, other translations render it more like:

> In the beginning, when God created the heavens and the earth—and the earth was without form or shape, with darkness over the abyss and a mighty wind sweeping over the waters—then God said: Let there be light, and there was light (Genesis 1:1–3 NAB)

The first suggests that the material world was created by God, resulting in a disorganized universe, which God then began to shape. The second suggests that the material world already existed when God started creating, and God was simply using that material. The dispute over which translation is correct gets into technical questions of Hebrew grammar that are beyond our scope. For our purposes, it is sufficient to note that ancient translations all translated it as a separate sentence, implying a prior act of creating the material universe. This is what the authors we will study would have been familiar with.

Much has been written on whether or not the doctrine of creation *ex nihilo* is taught by the Scriptures.[1] Some argue that it

---

[1] Chambers, *Reconsidering Creation Ex Nihilo in Genesis 1*; Ostler, "Out of Nothing"; Copan, "Creation Ex Nihilo or Ex Materia? A Critique of the Mormon Doctrine of Creation"; Copan and Craig, *Creation out of Nothing*.

is explicitly taught. Others argue that it is not explicitly taught but can be derived as an implication of other Biblical doctrines. In particular, if God alone is eternal, then matter cannot be eternal and thus must be created from nothing. Others argue that the text actually teaches creation from pre-existing materials and that creation *ex nihilo* is a foreign concept forced on the text.

There is a special challenge in interpreting both Biblical and extra-Biblical authors on this issue. Even very direct, explicit statements can be plausibly read in different ways. If an author claims that God created the world out of nothing, it is often debatable whether "nothing" means absolutely nothing or may instead refer to some sort of disorganized material that might, in some sense, also be described as nothing. Likewise, if an author claims that God created the world out of pre-existing material, it is debatable whether that material is understood to be eternal or simply created by God at a prior point in time. Accordingly, the views of authors, both Biblical and extra-Biblical, writing before the development of the doctrine of creation *ex nihilo* are subject to much debate.

However, creation is not the only event described in the Bible that violates the conservation of matter. In 2 Kings 4:1–7, Elisha oversees a miracle where one vessel of oil is used to fill a myriad of vessels. Jesus performed two miraculous feedings of a large number of people using a small amount of food. Both of these would seem to require the ex *nihilo* creation of new matter. Less obviously, this was probably also necessary for Jesus' miracle of transforming water into wine.

The Bible clearly indicates that God can violate the conservation of matter. It does not reject the conservation of matter in general. In fact, the violations are clearly seen as miraculous and something that only God could do. It only makes sense for violations of the conservation of matter to be miracles if the conservation of matter is generally true.

Historical Christians

Justin Martyr

Justin Martyr claims that Plato took his account of creation out of shapeless matter from the creation account of Moses.[2] This is a common theme in early Christian apologetics, which sought to show that any truth in Greek philosophy ultimately traced back to the Jewish prophets. He says that the whole world was made out of this formless matter by the word of God.

This passage is frequently cited to show that Justin Martyr held to creation *ex materia* and not creation *ex nihilo*. However, the passage makes no claim that creation *ex nihilo* disputes. God did create the world out of the waters, but God had previously created these waters. Justin Martyr does not explicitly claim that the waters were eternal or uncreated. Nevertheless, it may be argued that Martyr implicitly agreed with Plato that the matter was eternal by not explicitly disagreeing with him. It is safest to conclude that we simply do not know for certain what Justin Martyr thought the origin of the waters to be.

Tatian of Adiabene

Tatian was a pupil of Justin Martyr and one of the first to clearly express the doctrine of creation *ex nihilo*.[3] He states that once Jesus was begotten, he created matter and then used that matter to create the world. He compares this to his own action of trying to bring order to the confused matter around him. In context, he is referring to the words of his book bringing truth

---

[2] Martyr, "The First Apology," Chapter 59.
[3] Tatian, "Tatian's Address to the Greeks," Chapter 5.

to those who hear them. However, unlike Christ, he is bringing order to matter, which is his kindred, having also been created, and not that which he created himself.

Matter, he insists, is not eternal like God. It has a beginning and is thus not of equal power with God. Nor could any being besides God, the framer of all things, create matter. Thus, Tatian takes the position that God is the creator of matter, but that this is an ability restricted to God alone.

Origen of Alexandria

Origen mentions the idea that matter was itself uncreated.[4] However, he says that such a view cannot be held by Christians. He cites two passages from apocryphal works:

> I beg you, my child, to look at the heaven and the earth and see everything that is in them, and recognize that God did not make them out of things that existed. And in the same way the human race came into being. (2 Maccabees 7:28)

> No, said she; but hear the words which I am going to speak to you. God, who dwells in the heavens, and made out of nothing the things that exist, and multiplied and increased them on account of His holy Church, is angry with you for having sinned against me. (Shepherd of Hermas, Book 1, Vision 1, Chapter 1)

These works are both early indications of a belief in a creation from nothing. However, the meanings of these passages are disputed. In particular, it is disputed whether the passage intends to convey that God created the world from absolutely nothing or from some sort of formless matter, which might be considered nothing in a relative sense. Whatever the original

---

[4] Origen, "Commentary on the Gospel of John," Book I, Chapter 18.

authors' intention, Origen takes both passages as teaching creation *ex nihilo.*

Lactantius

Lactantius rejects the opinion of poets and philosophers who claim that God could not produce things out of nothing but rather required pre-existing materials.[5] These poets and philosophers indicated that God merely divided and arranged each object from that original, confused heap. Lactantius states that those who think this do not understand the power of God, believing that he cannot create something from nothing. He illustrates the argument by quoting Cicero, a Roman statesman and philosopher, and comparing the divine creation of the world to a builder making a building. Both, Cicero thinks, use materials but do not make the materials themselves. Cicero thus believed that matter, and thus the elements of earth, water, air, and fire, were not made by divine *fiat* but only used by divine powers. The quotation is presented as being from Cicero's book, *On the Nature of the Gods,* but it is not present in our extant copies of this work.

Cicero had claimed that it was improbable that God made matter, but Lactantius demands to know on what basis this conclusion is reached. Rather, Lactantius thinks it is highly probable. God is not limited by the weaknesses of man. If God requires the assistance of matter, how does the divine power differ from that of a mere man?

Cicero had also stated that it was probable that matter had always had a force and nature of its own. However, Lactantius argues that someone must have given it this force and nature. Where could it come from unless that source was God? It cannot be attributed to nature because nature lacks the under-

---

[5] Lactantius, "The Divine Institutes," Book II, Chapter IX.

standing to make something. Anything with the necessary foresight, skill, and power to create matter must be God.

Lactantius argues that matter and God cannot both be eternal. According to him, two eternal entities would necessarily be in conflict. One would overpower the other in this conflict, causing discord and destruction. Thus, there can only be one eternal entity, and all things must ultimately derive their origin from him.

Lactantius argues that any physical body can be influenced by an external force. Matter clearly has a physical body. But anything that can be influenced by an external force can be destroyed and therefore perish. Anything that can be destroyed or perish must also have an origin. Put another way, anything with an end must have a beginning. That which has a beginning must have a source. That source has to be an intelligent being, none other than God.

Athanasius of Alexandria

Athanasius describes the view of Plato as being that God made all things of pre-existent and uncreated matter.[6] This is probably not the view of Plato, who rather believed that time began at creation. However, the Neoplatonists had reinterpreted his creation account to make the world eternal in order to make his views agree with Aristotle. Athanasius argues that while a carpenter needs wood in order to create, that is a limitation of the carpenter. God, as the creator and not merely a craftsman, has no such limitation. Instead, he is able to bring raw matter itself into existence.

---

[6] Athanasius, "On the Incarnation of the Word," Chapter 1, Section 2.

Ambrose of Milan

Ambrose asks, if matter had no beginning, where was it prior to the creation of the world?[7] If one claims that it is located somewhere, one must think that not only matter but this place is eternal. Ambrose's implication is that such an eternal place seems too absurd to believe. Perhaps, Ambrose suggests, we ought to imagine this matter in the form of something like a flying earth. But in that case, the earth has no foundation, so what would keep it from falling? Perhaps it has wings? But where could such wings come from?

Ambrose suggests that we might interpret a couple of passages from Isaiah as indicating that the earth has wings:

> From the ends (wings) of the earth we hear songs of praise, of glory to the Righteous One. But I say, "I waste away, I waste away. Woe is me! For the traitors have betrayed, with betrayal the traitors have betrayed." (Isaiah 24:16 ESV)

The ESV translates the word as "ends," but the word literally refers to wings.

> Ah, land of whirring wings that is beyond the rivers of Cush, (Isaiah 18:1 ESV)

The intended meaning of "whirring wings" is unclear. It would be possible, but not likely, that this is referring to the earth's having actual wings.

However, if the earth had wings, where was the air it was flying through? Air did not exist because all matter was still unformed. Air, Ambrose points out, is not merely empty space but a physical substance. He supports this by referencing a passage from the Apocrypha:

---

[7] Ambrose of Milan, "Hexameron," Book 1, Chapter 7.

> Or as, when an arrow is shot at a target, the air thus divided, comes together at once, so that no one knows its pathways (Wisdom of Solomon 5:12 NRSV)

The passage indicates that air is a substance capable of being divided, rather than merely empty space.

Ambrose considers the suggestion that the material was located in God himself. However, he rejects this because God is a spirit outside the physical universe. As evidence, he references Paul:

> who alone has immortality, who dwells in unapproachable light, whom no one has ever seen or can see. To him be honor and eternal dominion. Amen. (1 Timothy 6:16 ESV)

He further references Jesus's words proclaiming that he and his servants were not "of this world." Therefore, the material for the world could not have come from God himself. Instead, God created the material for the world out of nothing.

John Chrysostom

Chrysostom, in response to the argument that creation from nothing was impossible, asks what the first man was made from, and indicates that everyone would agree that he was made out of the earth.[8] This is the Biblical account, as well as being present in many ancient mythologies. Chrysostom points out that nobody knew how you might create flesh out of earth. He mentions the many different parts of the body: bones, nerves, arteries, fat, skin, nails, and hair. Furthermore, even the bread we eat is changed into blood, gall, bile, and various other fluids. But bread has the color of grain, whereas blood is red or black (we would describe the blood in veins as appearing blue rather than black; the ancient Greek color classification

---

[8] Chrysostom, *Homilies on Genesis*, Homily 2, Section 11.

differed from our own). How can such a transformation be explained? If these critics cannot explain even the common transformations around us, what business do they have in declaring what God could and couldn't do? Chrysostom falls back on the Biblical text, insisting that the one sentence is sufficient to tear down all human reasoning and lead even these critics to the truth.

Chrysostom mentions those who claim that some sort of matter and darkness pre-existed God's creation of the world. He says that such a reading ignores the sequence of the text and pays no attention to the words of Moses. In his reading, the text first says that God created the heavens and the earth and then that the earth was "invisible and lacking all shape." This is how the Septuagint translated "formless and void." Chrysostom asks what could be worse than such madness and calls it idiocy. The creator is not human and does not require some sort of basis for his creation. Rather, he speaks, and it comes into being.

Basil of Caesarea

Basil insists that uncreated matter would be of equal rank with God.[9] He goes so far as to say that it would be wicked to suggest that formless and shapeless raw matter would be equal to the wise and powerful God. If the power of God was limited to acting on matter, this would make the power of matter equal to the power of God. Furthermore, this would be blasphemous because it proclaims that God is unable to finish his works on his own. Those who take this position do so because they are deceived by the limitations of human nature. All human crafts use materials such as iron or wood, but this does not apply to God.

---

[9] Basil of Caesarea, "Hexaemeron," Homily 2, Section 2.

Basil further elaborates on the opposing view. They believe that the form of the world comes from God, but matter comes from outside the world. There is thus a double origin for all things. Consequently, they believe God is not presiding over the origin or formation of the universe but only makes a final contribution to already ongoing work. For humans here on earth, the arts do require matter. Wool must exist before weaving, wood must exist before carpenters, etc. God is not so restricted. He imagined the world as it ought to be, and he created matter in pursuit of that goal. He created the elements as he wished. Basil rejects the idea that God created the form and not the matter. He creates both the essence (form) and the matter. He does not create just the figure but also the material that fills the figure.

Augustine of Hippo

Augustine describes the two-stage creation process.[10] God first made matter, then made everything else from that matter. He rejects the idea that God could not have made something from nothing. The opposing argument derives from the observation of carpenters, silversmiths, goldsmiths, and potters who need wood, silver, gold, or clay. They need the help of their materials. But the almighty God does not require any such help. In fact, it is sacrilegious to believe that he does.

John Philoponus

Philoponus argues against the idea that God could not create from nothing.[11] He grants, for the sake of argument, that in nature things are always created out of pre-existing materials. Nature exists and acts in the substrate of the material world.

---

[10] Augustine, "Two Books on Genesis against the Manichees," Book 1, Chapter 6.

[11] Philoponus, *Against Aristotle on the Eternity of the World*, Fragment 115.

Thus, it cannot act outside of the substrate and is thus unable to create matter. However, God has his own reality outside of nature and is not restricted to creating from pre-existing matter. If God were unable to do this, he would not be superior to matter. Thus, God produced not only the forms of all material things but even the matter they are made of.

Philoponus argues that if God creates matter and matter is not made up of some other material, then God can create something out of nothing. Matter might conceivably either be created all at once at the beginning of the universe or continuously throughout the history of the universe, but in either case, the matter does not require any sort of prior material to be created out of. This argument seems to beg the question, because Philoponus's opponent would not concede that God created matter. However, it does serve to show the internal consistency of the *ex nihilo* position.

The fact that objects in nature are always generated from existing things does not necessarily mean that the same is true for things created by God. Things formed in nature are produced through a gradual process. However, God simply brings things into existence through an exercise of the divine will. This is very different from the way things are formed in nature and is thus not necessarily subject to the same requirement for pre-existing material.

Remigius of Auxerre

Remigius describes the philosopher's account of the origin of the world as various "absurd opinions."[12] According to Plato, the demiurge, the being responsible for the creation of the world, formed the world out of matter based on the examples from the eternal world of the forms. Remigius refers to the

---

[12] Remigius of Auxerre, "Exposition on Genesis," Introduction.

demiurge as God and the forms as the exemplars. Plato views God as an artisan who shapes matter according to an eternal example. Remigius briefly mentions Aristotle as claiming that the world was eternal. However, Remigius dismisses both as in conflict with the words of Moses.

Thomas Aquinas

Aquinas addresses the objection that matter cannot be created because it is not made up of any other material.[13] Aristotle had argued that since there is nothing to make matter from, matter must not have been made at all but is instead eternal. Aquinas accepts Aristotle's logic in ordinary circumstances—normally, matter cannot be created or destroyed. However, he argues that this does not apply to special acts of creation, such as God's creation of the world.

Conclusions

The Christian Interaction

There are few issues on which the Christian response has been as consistent as that of creation *ex nihilo*. It may be that some very early theologians in the Church did not hold to the doctrine, but it has been the consistent position of the Christian Church since around the second century. Even among the diverse theological controversies of the early Church, this point was nearly universally accepted. As Origen stated on the subject:[14]

---

[13] Aquinas, *The Summa Theologiæ of St. Thomas Aquinas*, First Part, Question 46, Article 1.

[14] Origen, Cavadini, and Lubac, *On First Principles*, Book III, Chapter 5.

> Even the heretics, although widely opposed on many other things, yet on this appear to be at one, yielding to the authority of Scripture.

Christians accepted that the principle, nothing comes from nothing, was true for processes in nature. God, however, was not bound by this limitation. He could and did create matter from nothing. Chrysostom is a possible exception. He points out that the philosophers of his time could not explain even the ordinary transformation of bread into bodily fluids. He does not outright reject the principle that nothing comes from nothing in nature, but arguably seeks to undermine confidence in it.

The most common argument against creation from prior material was a citation of Genesis 1:1. This was widely taken as declaring that God created matter as the first step of creation. Another common argument was that God's almightiness implied that he could create matter out of nothing. To declare otherwise was to make God weak in the same way that man was weak.

There is a similarity between the general response to the conservation of matter and Aquinas's response to the immutability of the heavens. In both cases, the response accepts the principle as true for nature but insists that God is not bound by the same rules. However, while most of the authors rejected the immutability of the heavens even under natural processes, most accepted the conservation of matter under natural processes. Indeed, some of the same authors rejected the immutability of the heavens while accepting the conservation of matter.

Why the difference? In many cases, these authors explicitly point to Biblical passages that conflict with the immutability of the heavens. On the other hand, there are no Biblical passages that indicate that matter is not conserved under natural processes. Indeed, it may be argued that these passages imply the conservation of matter by claiming miraculous contravention of it. The difference is thus simple: according to the assessment of most historical Christians, the immutability of the heavens contradicted Scripture, but the conservation of matter did not.

Modern Science

Modern science inherited the idea of the conservation of matter but took some time to formalize it. The difficulty was that many processes involve either absorbing or releasing gases. It was not until techniques were developed to accurately weigh gases that it could be determined that all processes, even seemingly destructive ones like fire, did not change the total weight of those substances. Thus, it was determined that the total amount of matter, called "mass," in a closed system does not change. This is called the conservation of matter.

Einstein's theory of relativity refined the idea of the conservation of mass. He discovered that matter can be converted into energy and *vice versa*. This is the meaning of the equation, $E = mc^2$, which states that a small amount of matter is equivalent to a large amount of energy. Some common processes, like fire, do lose matter by converting it into energy, but the amount is so small that it cannot be measured. Nevertheless, while the conservation of matter is not strictly followed, the combination of mass and energy is conserved. As far as natural processes are concerned, nothing comes from nothing.

What does this mean for the origin of matter? The eighteenth century saw widespread acceptance of an eternal universe. On this understanding, the universe has always existed and has always looked essentially as it does today. Over time, stars and planets were formed and destroyed, but on a large scale, the universe fundamentally has always been approximately the same and always will be.

However, a couple of discoveries would overturn this theory. The first was the development of thermodynamics in the nineteenth century. Over time, entropy, essentially the amount of disorganization, of a closed system always increases. If matter and energy are not created, destroyed, or otherwise enter or leave the universe, it must be a closed system. Thus, if the universe were infinitely old, it would have infinite entropy and be maximally disorganized. If true, that would imply that stars, let alone life, could not exist.

The twentieth century saw the discovery of other galaxies. Furthermore, these galaxies appear to be moving away from us. Today, we understand this to be the universe expanding. However, if the universe had been expanding for an infinite amount of time, these galaxies ought to be infinitely far away.

The combination of these two discoveries suggested that the universe had a finite history and not an eternal one. The most famous attempt to avoid this conclusion was the steady-state model most often associated with Fred Hoyle. This theory postulated the continual creation of matter. This allowed it to avoid the problem of entropy because the universe is no longer closed. It also resolved the problem of the expanding universe because it allowed new galaxies to continually form to replace those that have moved further away. The theory only required sacrificing the principle of the conservation of matter in order to save an eternal universe.

Today, almost all cosmologists accept the Big Bang theory. This theory postulates that the universe as we know it expanded from a hot, dense state around fourteen billion years ago. The universe, as we know it, has a finite history or age. It has not existed eternally and has changed in significant ways over time.

The Big Bang is highly suggestive of an *ex nihilo* creation event that created our universe. However, it is still possible to avoid this implication by postulating a history prior to the Big Bang. The Big Bang theory only involves the development of the universe from the "singularity" (the point where the equations break down), not the origin of that singularity. For example, oscillating theories postulate that the singularity resulted from the termination of the last one in an eternal sequence of universes. The chief difficulty for any such theory is entropy; any theory postulating an eternal history of the universe has to explain why the universe is not already at maximal entropy.

Modern science has established two important laws with seemingly opposite implications for the eternality of the universe. On the one hand, modern science has confirmed the conservation of matter and energy; matter is not created or destroyed but only changes forms. But if matter cannot be created or destroyed, this implies that it must have simply always existed. The universe must then be eternal. On the other hand, science has established the principle of increasing entropy. Over time, the entropy of the universe is increasing, and if it were eternal, it ought to be at maximal entropy. Since the universe is not at maximum entropy, this implies that it cannot be eternal. Thus, these two laws, while well established scientifically, have conflicting implications.

Evaluation of the Response

Historically, Christians have accepted the conservation of matter while insisting that God could nevertheless create matter. This position resolves the conflict between the conservation of matter, which suggests that the universal is eternal, and entropy, which suggests that the universe cannot be eternal. God created the material universe from nothing at some point in the finite past. This miraculous event explains the origin of matter, thus explaining the apparent conflict between these two theories. Outside of miraculous interventions, the total amount of matter and energy has remained constant while entropy has increased since then.

Indeed, modern science's acceptance of a universe with a finite age has made it easier to defend creation *ex nihilo* than at any previous time in history. The historical Christian defended creation despite the lack of clear scientific evidence that the universe was not eternal. Today, those defending an eternal universe must deal with the clear scientific evidence suggesting that the universe has a finite history. It is still possible to defend an eternal universe, but it is much more difficult. This has made pointing to this evidence, as a clear reason to infer the existence of God, a staple of modern Christian apologetics.

Historical Christians could have responded to this issue by reconciling creation with eternal matter. They could have decided that God had made the universe out of uncreated, eternal matter. Indeed, we saw that there are Biblical passages that describe creation out of existing matter. Joseph Smith, the founder of Mormonism, did reconcile this, incorporating eternal matter into Mormon theology. However, Christians refused to do this, insisting that the Scriptures taught the creation of the world from nothing.

On the other hand, they could have responded by attempting to discredit the principle of the conservation of matter. They could have argued against the idea that matter was not created or destroyed. Perhaps they could have argued that fire really did destroy what it consumed and did not merely transform it. Nevertheless, they did not make this argument but instead accepted that the conservation of matter was true.

Instead, Christians took the position that would best align with modern science. Modern science has, in some sense, vindicated the historical Christian beliefs on this issue. This was not lost on modern critics of Christianity, some of whom have gone to great lengths to avoid the theistic implications of modern scientific discoveries.[15]

We see again a point of conflict between the science of the historical day, which claimed that nothing came from nothing, and the claims of Scripture. Historical Christians consistently followed the authority of Scripture and rejected the scientific claim. By reevaluating the scientific argument, Christians saw that the conservation of matter did not imply the eternality of matter.

---

[15] Craig and Smith, *Theism, Atheism, and Big Bang Cosmology.*

Chapter 4

Has Mankind Always Existed?

The Scientific Question

Which came first, the chicken or the egg? If you know anything about chickens, you know that every chicken hatches from an egg. But you are also aware that every egg was laid by a chicken. We know these things from observation, and nobody has ever observed an exception to these rules. However, if these rules are taken to be absolute, the inevitable conclusion is that there was no first chicken. A first chicken would have had to come from an egg, and therefore from a previous chicken that laid that egg. Instead, there must be a lineage of chickens and eggs stretching back into the infinite past.

But this logic is not restricted to chickens. It applies to all lifeforms because all lifeforms follow the law of biogenesis, which states that living things arise only by reproduction from prior living things. In the ancient world, this was not accepted as an absolute principle; instead, they believed there were exceptions. They believed that some living things arose by spontaneous generation without biological precursors. Nevertheless, they believed that many forms of life did follow the law of biogenesis.

In particular, humans followed this law. Every human has a mother who gave birth to them. But those mothers, in turn, have their own mothers. Indeed, no human has ever been observed to arrive in any other way than by birth by a woman. No human ever arises by spontaneous generation. But if this rule is absolute, there cannot have been a first human, just as there cannot have been a first chicken. Instead, there must be an infinite lineage of humans tracing back in history. Humans have simply always been here.

Today, the idea that chickenkind or mankind could be eternal seems absurd. It would never cross our minds to seriously consider the possibility that either chickens or men have existed for all eternity. However, in the classical world, where many argued that the heavens and matter were eternal, this did not seem so far-fetched. Rather, the ideas that matter existed eternally, the heavens were immutable and eternal, and mankind had no beginning fit together as a plausible understanding of the world.

History, however, poses a problem for this understanding. If mankind has been exploring the earth for an eternity, nobody should be able to discover any new lands, because every land must have already been discovered in that eternal past. If artisans have been studying their craft for an eternity, nobody should be able to improve upon what has already been discovered. All technological innovations ought to have already been found. All mathematical and scientific discoveries ought to have already been made. Fundamentally, with an infinite past, there should be no firsts, either now or in recorded history.

However, this is not what we find in history. Instead, we have people discovering new lands and populating them. New technologies, discoveries, and arts are developed. History is replete with firsts; it does not look like the history at the end of an eternity of human civilization. Rather, it looks like the history we would expect if humans were relatively recent arrivals.

Furthermore, there is not enough recorded history if humans have been around for an eternity. There ought to be eons upon eons of human civilizations stretching back in time. We know of ancient civilizations such as Egypt, Mesopotamia, and China. But if humans have been living on earth for an eternity, there ought to be civilizations more ancient still. There ought, in fact, to be infinite civilizations going back in time. But these civilizations are conspicuously absent.

There was, however, a proposed explanation for these discrepancies. It was claimed that various cataclysms, such as floods and fires, would occasionally destroy human civilization. Any humans who survived would be reduced to a primitive state. They would forget the history, arts, and discoveries of the past civilization. Thus, human civilization would begin anew.

Consequently, the various firsts that are recorded in history were not actually firsts. Rather, they are merely the first since the latest reset of civilization. These firsts have happened over and over again during the infinite history of humanity. History seems short because we only know the history back to the most recent cataclysm.

This idea had some support from certain nations that claimed to have recorded long histories. One example is the Sumerian Kings List, which purports to describe the reigns of kings, some of whom reigned for tens of thousands of years, making up hundreds of thousands of years of claimed history. Another example is the now-lost book *On The Gods in Egypt*, which contained an apocryphal letter from Alexander the Great claiming tens of thousands of years of Egyptian history. A similar story appears in Plato's *Timaeus*, in which an Egyptian priest relays history that was lost to the Greeks due to previous cataclysms. Both the Babylonians, heirs of the Sumerians, and the Egyptians claimed long histories beyond those known by the Greeks or other cultures. These accounts, it was claimed, provided evidence of past history, which had been mostly lost in cataclysmic events.

The Biblical Discussion

The previous chapters have discussed the Biblical case that the heavens and matter itself are not eternal. Naturally, if the world itself is not eternal, mankind is also obviously not eternal. Additionally, the Bible indicates that God created the first man and woman, directly implying that mankind is not eternal.

> So God created man in his own image, in the image of God he created him; male and female he created them. (Genesis 1:27 ESV)

> He answered, "Have you not read that he who created them from the beginning made them male and female, and said, 'Therefore a man shall leave his father and his mother and hold fast to his wife, and the two shall become one flesh'? (Matthew 19:4–5 ESV)

> And he made from one man every nation of mankind to live on all the face of the earth, having determined allotted periods and the boundaries of their dwelling place, (Acts 17:26 ESV)

A strong Biblical case can be made that the Bible teaches not only the temporal origin of humanity but also its recent origin. First, Adam is explicitly identified as the first man.

> Thus it is written, "The first man Adam became a living being"; the last Adam became a life-giving spirit. (1 Corinthians 15:45 ESV)

Secondly, three different Biblical genealogies indicate that Adam lived a relatively small number of generations ago. Genesis 5 and 11 give a genealogy from Adam to Abraham. 1 Chronicles 1–2 gives a genealogy from Adam to David. Luke 3:23–38 gives a genealogy from Adam to Jesus Christ. Chronological information, especially in Genesis 5 and 11, can be used to estimate how long ago Adam lived. However, even without that information, the number of generations given since Adam implies that he lived within the past several thousand years, even allowing for some gaps in the genealogies.

There are modern Christians who argue that the Bible is consistent with humans' having existed for much longer than several thousand years. However, they do not hold to a unified position that could be readily described, defended, or critiqued. Rather, they hold to a diverse set of views that attempt to deal with the Biblical case for a recent origin of humanity in a variety of ways. It would be impossible to fairly characterize these diverse views here.

For our purposes, it is sufficient to appreciate the weight of the Biblical case for the recent origin of humanity. Even those who dispute it must admit that it is not readily dismissed. It is not based on unusual exegesis or mistranslated texts. We can at least understand why historical Christians understood the Bible to be definitively claiming such a recent origin.

The issue of the recent origin of humanity should not be confused with the issue of the age of the earth. It is conceivable that humanity is young while the earth is old. Indeed, many advocates of old-earth creationism believe this to be the case. Our focus in this chapter is on the origin of humans, and we will not further discuss the question of the age of the earth.

Historical Christians

Origen of Alexandria

Origen discusses this issue in response to Celsus.[1] According to Celsus, the flood of Deucalion, the Greek counterpart to Noah, was merely the most recent cataclysm in a series of cataclysmic floods and fires. Origen insists that this reveals a "secret desire" to discredit the Mosaic creation account and instead claim that the world was eternal. In contrast, Origen appeals to the Mosaic account of creation, which indicates that the world is much less than ten thousand years old.

Origen's first response is to simply demand an argument to support the claim that these cataclysms took place. He refuses to accept the unsupported assertion that these prior cataclysms occurred. He mentions, as before, the flood of Deucalion. Additionally, he mentions the conflagration of Phaëthon. In Greek mythology, Phaëthon was the son of the sun god, Helios, who drove his father's chariot, the sun. However, he was not able to control the horses and so burned the earth. Both the flood and fire were cataclysmic events, according to Celsus, just the most recent of an infinite series of such events.

Origen anticipates a reference to Plato's dialogues to support the claim that there have been many of these cataclysms. In particular, he is referring to a passage from *Timaeus*, from which we derive the story of Atlantis. According to Plato's account, Atlantis was an island nation that had invaded Greece nine thousand years previously but was then defeated by the city of Athens. But this had been forgotten by the Greeks because a flood had destroyed their civilization between that event and Plato's day. In the *Timaeus*, this story is relayed by an

---

[1] Origen, "Contra Celsum," Book 1, Chapter 19.

Egyptian priest who explains that Egypt was protected from these fires and floods by the special properties of the Nile.

Whether or not Plato thought he was conveying an accurate historical account is a matter of debate. Most scholars today believe that Plato was simply telling the story as an allegory. He did not intend to imply that there was ever an Atlantis or even an Egyptian priest relaying the story of Atlantis. Nevertheless, many in the ancient world did take it as purported history, and this is how Origen understands it.

Origen responds to this possible argument by insisting that Moses provides more reliable information than Plato. He speaks of Moses as being pure, pious, ascending above all created things, and uniting himself to the Creator. He is, according to Origen, better than all of the wise men who lived in the Greek or Roman civilizations. If Celsus demands a reason for this assessment, Origen demands that Celsus first provide reasons for his position.

Origen argues that Celsus testifies to the youth of the world because he acknowledges that the Greeks understood the flood and conflagration as ancient events, being unaware of any prior events. This in and of itself suggests that there were no prior events. Therefore, Origen argues, Celsus is effectively confirming the recent origin of humans.

The story in *Timaeus* was attributed to an Egyptian priest. Origen seeks to undermine the reliability of those Egyptian priests and, thus, these stories. He points out that the Egyptian priests worshiped irrational animals. Why should we accept the testimony of the Egyptian priests when they clearly believe absurd things?

Augustine of Hippo

Augustine discusses the idea that mankind is eternal.[2] He quotes a Platonist philosopher and novelist, Apuleius, describing man as mortal but mankind as immortal. For these philosophers, this idea is supported by appealing to the observation that man only arises by being produced by man. However, Augustine makes the counterargument that if mankind is eternal, how can recorded histories be accurate? In an eternal world, there should not be inventors, institutors, explorers, or colonists. However, Augustine explains that some reply that there are periodic destructions of all lands by fires and floods. Thus, there are periodic new beginnings that reset history. Augustine dismisses this as what they *think* rather than what they know. That is, this is just speculation without any support.

Augustine indicates that some accept these repeated cataclysms and long histories on the basis of "highly mendacious documents." These documents claimed to give accounts of histories spanning many thousands of years beyond that of the traditional Christian timescale. He cites a particular example: a letter purportedly written by Alexander the Great to his mother. This letter appeared in a book, *On the Gods of Egypt*, by Leon of Pella who lived in the 4th century BC. We know little about the letter, book, or author. What we do know is that the book argued that the Egyptian gods were actually ancient humans whose exploits had been mythologized. The early Christians often argued the same and cited this work to support their claim. The letter, now understood to be apocryphal, purports to relay information from an Egyptian priest regarding these past events. In particular, it purported to record thousands of years of history beyond that of the traditional Christian timescale.

---

[2] Augustine, "The City of God," Book 12, Chapter 10.

Augustine points out that this letter assigns time periods to ancient kingdoms, which differed greatly from the Greek accounts. In particular, he speaks of the Assyrians, Persians, and Macedonians. The chronology presented in the letter had these kingdoms lasting for thousands of years, whereas Greek history only attributed hundreds of years to those kingdoms. Modern historians give numbers approximately in line with those attributed to the Greeks here. Augustine attributes this to the Egyptians counting four months as a year. Thus, there would have been three Egyptian years for one regular year. The Egyptians did have a three-season calendar system based on the flooding of the Nile. However, we do not have corroboration for the idea that these seasons became confused with years. Furthermore, as Augustine points out, even with such a multiplication, the numbers do not fit Greek history. He insists that the Greek numbers must be given greater credit because they agree with the Biblical text.

Furthermore, Augustine argues that if the letter of Alexander the Great cannot be trusted, this is even more true of other accounts. These accounts are full of fabulous and fictitious stories. It makes no sense to put these documents against the authority of the Scriptures. Rather, the Scriptures are to be trusted because of their successful track record of predictions. In particular, he points to the successful prediction that the Bible would be believed by the whole world. Obviously, the whole world did not believe, but Augustine probably had in mind the Roman Empire, which had largely converted to Christianity.

Cosmas Indicopleustes

Cosmas argues against the eternity of mankind.[3] He argues that the arts and technologies gradually develop over time. Gradually, the arts are brought to greater perfection through practice. He turns this into an argument against the eternity of mankind. If arts and technologies developed gradually over time, then if we went far enough back in time, we would find people without any arts or technologies. They would be unable to weave clothing, build houses, cure diseases, or reap crops. How could they have survived? Rather, it must be the case that the world is not eternal but rather was created recently.

Thomas Aquinas

Aquinas argues that the eternity of the world cannot be known except by special revelation.[4] That is, we know that the world was created rather than eternal because this is revealed in God's word. However, he argues that no philosophical or scientific argument can prove this. Aquinas considers the argument that arts have developed and countries have become inhabited over time, and thus the world cannot be eternal.[5] He points out that those who hold to the eternity of the world can simply argue for cycles of progress and decay.

Certain countries become inhabitable and uninhabitable at different times. Likewise, the arts are developed and lost over time. Aquinas quotes Aristotle, declaring that it is absurd to conclude that the world is new because of this argument.

---

[3] Indicopleustes, *Christian Topography*, Book III.

[4] Aquinas, *The Summa Theologiæ of St. Thomas Aquinas*, First Part, Question 46, Article 2.

[5] Aquinas, First Part, Question 46, Article 2, Response To Objection 4.

Conclusions

The Christian Interaction

All of these authors agreed that mankind was not eternal. This is a necessary consequence of believing that the world itself is not eternal. Origen and Augustine explicitly claim that the world has only existed for several thousand years. This was in opposition to purported histories claiming a much longer period of time.

With respect to the historical argument against the eternity of mankind, there is some division. Three authors, Origen, Augustine, and Cosmas, make this argument against the eternity of mankind. Aquinas, on the other hand, rejects it as invalid, pointing out that it was at least logically possible that human civilization was simply periodically reset. However, Origen and Augustine both discuss this possibility, and reject it as baseless speculation.

Both Origen and Augustine considered books that claimed to record some of this more extensive history. They seek to undermine these accounts. Origen points out that the Egyptian priests from whom these accounts are supposed to derive believed in other nonsense. Augustine pointed out that the accounts differed from those of the Greeks. Furthermore, many of the accounts contain fantastical elements. As such, they insisted that the Scriptures were the more reliable guide.

Modern Science

The modern sciences of geology and paleontology began in the seventeenth and eighteenth centuries. These sciences revealed that earth itself had a history, filled with fossils and geological formations. According to the mainstream scientific interpretation, this was a history of long ages of time, and humans only came into that history at the very end. They are not eternal occupants of earth, but relatively recent arrivals.

Recorded human history traces back several thousand years. The claims of certain ancient civilizations dating back hundreds of thousands of years were unfounded. Instead, the oldest civilizations are dated back to the fourth millennium BC. However, if the standard account of the geological column is accepted, humans have been around for much longer, between hundreds of thousands and millions of years. Humans living prior to several thousand years ago were too culturally and technologically primitive to record history and build great civilizations. This would make humans relative newcomers to the whole geological column but much older than the several thousand years found in recorded history.

Evaluation of the Response

If we focus strictly on the question, "Is mankind eternal?," then modern science and the historical Christian are in agreement. Neither mankind in particular nor life in general is eternal but came into existence at some point in time. The timing and nature of that origin may be disputed, but the occurrence of that origin is not. This is the primary issue being disputed by the authors in this chapter. While other issues are involved, the central issue is the eternality of mankind, and on that issue, historical Christianity and modern science agree. On this point, the Christian claim has been vindicated by modern science.

However, Aquinas differed from the other Christian authors. While he accepted the Biblical claim that mankind was not eternal, he thought this could not be proven but was an article of faith known only through revelation. As he put it:[6]

---

[6] Aquinas, Part 1, Question 46, Article 2.

Hence it cannot be demonstrated that man, or heaven, or a stone were not always.

However, the evidence from modern science is very strong that mankind, heaven, and stones are not eternal. Even if the material world may be eternal, these things are clearly not.

There are two possible defenses of Aquinas. It may be argued that he was correct in his time that the eternity of the world was a matter of faith, but that this has since changed. On the other hand, it may be argued that modern science has not "proven" the non-eternality of the world but has only shown it to be highly probable. Nevertheless, it remains that Aquinas took a position defending the viability of holding to an eternal world and mankind against other Christian authors who had disputed it. Aquinas's position has not aged well.

Agreement between modern science and the historical Christian can also be found in the evaluation of the claims about history they were disputing. In particular, they disputed claims about tens, if not hundreds, of thousands of years of the history of civilizations. They also disputed the occurrence of repeated cataclysms that reset human civilization. On both of these points, they were correct. Ancient civilizations did not have these long histories, and human civilizations had not been reset by various cataclysms. Rather, recorded human history only goes back several thousand years.

However, historical Christianity and mainstream modern science do not agree as to why recorded human history only traces back several thousand years. According to the historical Christian, this was because there was no more history. According to mainstream modern science, there were no prior civilizations or recorded human history because humans were too technologically primitive. They did not have writing to record history or the technology to build great civilizations.

Furthermore, historical Christianity and mainstream modern science do not agree as to how long humans have existed. According to mainstream science, humans have existed on earth for hundreds of thousands to millions of years. The historical Christian held that humans had existed for a mere several thousand years.

The historical Christian chronology is harder to defend today than it was when these authors wrote. Even those who find the scientific arguments against the mainstream position compelling must acknowledge that they are better supported than the alleged histories our spiritual forefathers rejected. Regardless of whether or not the traditional Christian chronology is correct, it must be acknowledged that mainstream modern science has not vindicated the Christian claim but rather made it more difficult to maintain.

We must then ask if there is an identifiable error that led them to draw an incorrect conclusion. There is no obvious good candidate. The conclusion was not based on obviously specious exegesis. It was not a matter of reading the science of their day into the text. It did not depend on a faulty translation of the text. It seems difficult to argue that there was any way that the historical Christian should have seen that this was not the correct interpretation.

Again, we have a conflict between the Scriptural claim of a historical creation of man and the idea of an eternal mankind. Christians were correct to insist that mankind was not eternal. However, the traditional Christian chronology has not been vindicated by modern science. This issue remains with us today, albeit in a modified form.

Chapter 5

Can Formless Matter Exist?

The Scientific Question

What, fundamentally, is the world made of? Many things are composite; they are made up of a combination of other things. For example, a house is made up of wood, wires, pipes, etc. These are in turn made up of substances such as metal, plastic, and cellulose. But what are metal, plastic, and cellulose made of? What, at the fundamental level, is everything made of?

According to classical science, there were four elements making up all terrestrial things: earth, water, air, and fire. These were not quite the earth, water, air, and fire with which we are familiar, but rather idealized and pure versions of them. According to classical science, each of the elements had particular properties or qualities. Fire and air were both hot, as opposed to earth and water, which were cold. Water and air were both wet, whereas earth and fire were dry. Earth was solid, water was liquid, and air was gaseous. Classical science attempted to explain the properties of various things based on the properties of their constituent elements.

However, in the classical conception, elements were not immutable but could change from one into the other. For example, when water evaporated, it was thought to change into air, but turned back into water when it rained. It seemed that in order for such a change to be possible, there must be an underlying material, or matter, that these different elements were made out of. When one element changed into another, the form or shape changed, but the matter or substance remained the same. This would be akin to the idea that if a golden statue of a rabbit were melted down and reformed into a duck, the form would have changed, but the substance of gold would be the same. This only works because there is an underlying material common to both the original rabbit statue and the new duck statue.

This gave rise to a question: could formless matter exist? Was it possible for raw matter to exist apart from being in the form of any of the elements? Plato, in his book *Timaeus*, gives an account of the creation of the world. In that book, he describes space or matter as an invisible and formless being that could receive any form put on it:[1]

> In the same way space or matter is neither earth nor fire nor air nor water, but an invisible and formless being which receives all things, and in an incomprehensible manner partakes of the intelligible.

In Plato's account, this invisible and formless matter is shaped into the elements and world we know by the demiurge.

In contrast, Aristotle argued that matter and form were inseparable and that one could not exist without the other. Indeed, it is difficult to comprehend what it would mean for matter to exist without a form. One can easily envision matter existing as some sort of amorphous blob, but even an amorphous blob is still a shape or form. To exist utterly without shape makes little intuitive sense.

---

[1] Plato, *Timaeus*, Section 1.

The Biblical Discussion

Many modern English Bible translations appear to indicate that the earth was created without form or was formless:

> The earth was without form and void, and darkness was over the face of the deep. And the Spirit of God was hovering over the face of the waters. (Genesis 1:2 ESV)

However, while most modern English translations refer to earth as being without form, the Hebrew does not. Instead, it says that the earth was תֹהוּ (*tohuw*) and בֹהוּ (*bohuw*). *Tohuw* refers to emptiness and is used to refer to wildernesses, vanity, and empty places. *Bohuw* only appears in this verse and references to it. It is typically thought to be a word invented simply to rhyme with *tohuw*. The phrase "*tohuw* and *bohuw*" (*tohu wa-bohu*) simply refers to a disorganized state and not matter without any form whatsoever.

A straightforward reading of the text of Genesis 1 suggests that the whole world was in a watery state. Waters are mentioned at the beginning of the creation account, and everything else was created either by dividing those waters or emerging from those waters. The waters are divided into the waters above and the waters below. The waters below are divided into the seas and the dry land. Birds and sea creatures emerge from the sea. The land, which came out of the waters, is commanded to bring forth plants and animals. Man, like the animals, is created from the dust of the ground. Ultimately, all things trace their origins to those primeval waters.

Furthermore, 2 Peter possibly refers to the earth being created from these waters:

> For they deliberately overlook this fact, that the heavens existed long ago, and the earth was formed out of water and through water by the word of God, and that by means of

these the world that then existed was deluged with water and perished. (2 Peter 3:5–6ESV)

Clearly, Peter is referring to some relationship between the earth and the waters, but the exact relationship is uncertain. It is not entirely clear how to understand the phrase "out of water and through water." While the English Standard Version says "the earth was formed out of water and through water," the New King James Version says "the earth standing out of water and in the water," and the American Standard Version says "an earth compacted out of water and amidst water." These different translations reflect the uncertainty of the meaning of the Hebrew text of this verse.

A plausible reading is that Peter is retelling the creation of the land from Genesis 1:

And God said, "Let the waters under the heavens be gathered together into one place, and let the dry land appear." And it was so. God called the dry land Earth, and the waters that were gathered together he called Seas. And God saw that it was good. (Genesis 1:9–10 ESV)

The verb that Peter uses, συνίστημι (synistēmi), typically refers not to forming anything but rather to placing things together. It is a natural choice of verb to describe moving the land out of the water, as described in Genesis 1. The dry land was previously in the waters; it was thus taken out of the waters, which is why Peter says that the earth was formed "out of water." When Peter indicates that the earth is "through the water," he means that it is spatially through or in the midst of the waters, in the same way that the Nile runs through Egypt. Finally, Peter proclaims that everything took place "by the word of God," echoing Genesis 1:9, which describes God creating by simply giving a command. Altogether, the passage appears to closely correspond with the creation account of the dry land. Furthermore, this explains the reference to the flood: the waters that deluged the earth are those that the earth was taken out of and that surround the earth.

The Genesis 1 and 2 Peter accounts both appear to indicate creation from a watery, chaotic, and disorganized state. However, the Septuagint gives a rather curious translation:

> Yet the earth was invisible and unformed, and darkness was over the abyss, and a divine wind was carried along over the water (Genesis 1:2 New English Translation of the Septuagint)

This translation states that the earth was invisible and unformed. This is a curious translation, as the Hebrew does not suggest invisibility. Nevertheless, Plato is a plausible source for this idea. Plato described the initial formless matter in his conception of creation as "an invisible and formless being." Plato had lived a couple of centuries before the translation of the Septuagint, and it is widely thought that his ideas influenced that translation.

The Greek in the Septuagint word here, translated into English as unformed, ἀκατασκεύαστος (akataskeúastos), is of debated meaning. The 1851 Brenton translation of the Septuagint into English uses the word "unfurnished." The Orthodox Study Bible, also translated from the Septuagint, says "unfinished." Furthermore, we will find that many of the authors using this translation understand it as unfinished or unfurnished. As such, the Septuagint does not explicitly refer to formless matter, even if some modern translations suggest that it does.

However, the Septuagint does indicate that the earth was invisible. But what exactly does it mean for the earth to be invisible? Biblically, it is God and spiritual things that are described as invisible:

> For his invisible attributes, namely, his eternal power and divine nature, have been clearly perceived, ever since the creation of the world, in the things that have been made. So they are without excuse. (Romans 1:20 ESV)

> He is the image of the invisible God, the firstborn of all creation. For by him all things were created, in heaven and on earth, visible and invisible, whether thrones or dominions or rulers or authorities—all things were created through him and for him. (Colossians 1:15–16 ESV)

> To the King of the ages, immortal, invisible, the only God, be honor and glory forever and ever. Amen. (1 Timothy 1:17 ESV)

By invisible, the Scriptures mean that God is inaccessible to all of our senses. It is not that God cannot be seen but can be smelled, heard, or tasted. He is not part of the visible or physical world but of the invisible or spiritual world.

However, this is not the only possible meaning of invisible. The Greek word translated invisible literally just means unseen. An object might not be innately invisible but rather simply hidden from sight. The earth, many proposed, was hidden by being covered by the waters until the third day of creation:

> And God said, "Let the waters under the heavens be gathered together into one place, and let the dry land appear." And it was so. (Genesis 1:9 ESV)

The dry land is not said to have been created on the third day, but rather to have appeared as a consequence of the waters being gathered together. This would imply that the earth was previously covered by water and thus hidden from view. In other words, it was previously invisible.

Early Latin translations of the Septuagint describe the earth as *invisibilis* and *rudis*. *Invisibilis* simply means invisible. *Rudis* means rough, raw, or uncultivated. It roughly conveys the same notion of being unfinished or unfurnished as the Septuagint.

The Vulgate corrected the oddity of the Septuagint, removing the reference to an invisible earth:

> And the earth was void and empty, and darkness was upon the face of the deep; and the spirit of God moved over the waters. (Gen 1:2 Douay–Rheims Bible)

The Vulgate neither suggests the invisibility of the earth nor its formlessness.

This leaves us to wonder: how did the notion of the earth being without form make its way into modern English translations? Indeed, the earliest English translations do not describe the earth as being without form. The Wycliffe Bible, translated into English in the 14th century, described the earth as "idel," meaning empty, useless, or worthless, and "voide." William Tyndale's translation, produced in the early sixteenth century, describes the earth as "voyde" and "emptie." However, the Geneva Bible, translated under the influence of John Calvin in the late sixteenth century, described the earth as "without form" and "void". This translation was highly influential in its day and was followed on this point by the King James Version, which remains influential today. Many modern English translations give a similar reading, following this tradition.

A possible reference to formless matter is found in the book of Hebrews:

> By faith we understand that the universe was created by the word of God, so that what is seen was not made out of things that are visible. (Hebrews 11:3 ESV)

This is often interpreted as a reference to creation *ex nihilo*. But the text does not actually say that the universe was created out of nothing. Instead, it says that the universe was not made out of visible things. This would seem to suggest that it was made out of invisible things. What invisible thing might the world be made out of? Perhaps the author of Hebrews has in mind the invisible earth of Genesis 1:2 in the Septuagint and is endorsing the idea of creation from formless matter.

However, a closer look at the context suggests that this passage is not commenting on the material of creation at all. The entire passage is about men and women of faith in the Old Covenant, and a statement about the creation of the world seems rather out of place. The passage defines faith as the "conviction of things not seen," or, put another way, the conviction of invisible things. Throughout the passage, Biblical saints have faith in God regarding events they cannot see. Within that context, what is invisible is not formless matter but rather God and his actions.

Furthermore, the word translated as universe is αἰῶνας (aiōnas), which means ages. The word translated as formed is κατηρτίσθαι (katērtisthai), which refers to preparing or completing something. As such, the text speaks of preparing the ages, not forming the universe. Within the context, it appears to be referring to God's providential control over history and has nothing to do with the physical formation of the universe.

A text from the Apocrypha explicitly refers to formless matter.

> For your all-powerful hand, which created the world out of formless matter, did not lack the means to send upon them a multitude of bears, or bold lions (Wisdom of Solomon 11:17 NRSV)

The context of this verse is a discussion of an episode recorded in Numbers 21 during the journey to the Promised Land when the Israelites were attacked by serpents. The point of the verse is that if God could create the world out of formless matter, he certainly could have created bears and lions to punish those who rejected him. Despite its title, the Wisdom of Solomon was not written by Solomon but rather by a Greek-speaking Jew around the time of Christ. The word translated as formless in this text is ἀμόρφου (amorphou) which means shapeless, formless, or amorphous.

Historical Christians

Justin Martyr

Justin Martyr reads Genesis 1:1–3 as teaching that the earth was made from shapeless matter.[2] However, while he quotes Genesis, he does not elaborate on his exegesis. He argues that this is in agreement with Plato, going so far as to argue that Plato got his ideas from Moses. He supports this claim by pointing out that Moses wrote earlier than the Greek writers and, thus, by implication, must have been the original source.

Tertullian

Tertullian explains that God first created the uncultivated elements and then arranged them to give them finished beauty.[3] Between the creation of the earth and its perfection, it was invisible and unfinished. He explains that the earth was unfinished because it was invisible, and it was invisible because it was covered with water. To support this, he appeals to the fact that the text describes the earth as appearing, not being

---

[2] Martyr, "The First Apology," Chapter 59.
[3] Tertullian, "Against Hermogenes," Chapter 29.

created, when the waters were gathered together. Tertullian does not understand the text as referring to formless matter. Nor does he show any awareness of the notion of formless matter.

Origen of Alexandria

Origen addresses the issue of formless matter.[4] He begins by arguing that matter undergoes transformation from one element to another. He gives various examples. Wood is transformed into fire, then smoke, and then air. Oil is transformed into fire. Food is transformed into the substance of our bodies. He defines matter as the substance underlying these different elements, i.e., what those elements are made of. The different elements have different qualities: heat, cold, dryness, and humidity. When matter is given a form, the latter's qualities are implanted in that matter. Matter itself has no qualities, but it never exists as simply formless matter without such qualities. In other words, Origen argues that formless matter cannot exist.

Later, Origen argues that the Scriptures never use the word "matter" in the technical sense of that which underlies the physical elements.[5] He considers a verse in Isaiah that uses the same word but in a different sense:

> The light of Israel will become a fire, and his Holy One a flame, and it will burn and devour his thorns and briers in one day. (Isaiah 10:17 ESV)

In context, Isaiah is proclaiming judgment against Assyria. The Assyrian army is being compared to a forest that God will burn down. The Septuagint renders this as:

---

[4] Origen, Cavadini, and Lubac, *On First Principles*, Book II, Chapter 1, Paragraph 4.

[5] Origen, Cavadini, and Lubac, Book IV, Section 33.

> The light of Israel will become a fire, and it will sanctify him
> with a burning fire and devour the wood (matter) like grass.
> (Isaiah 10:17 New English Translation of the Septuagint)

The Hebrew refers to thorns and briars, which were translated using the Greek word for wood, ὕλη (hylē). Greek philosophers used the same word for wood and matter. Today, we often borrow words from Greek or Latin for specialized meanings. However, the Greeks had no ancient language to borrow from and thus often reused ordinary words, endowing them with a specific technical sense.

Undoubtedly, Origen is correct that this does not refer to matter in that technical sense. Origen interprets the matter as referring to sins, following his tendency towards allegorical interpretations. However, it is actually sufficient to understand the text as referring to simple wood.

Origen suggests that if the word does appear in any other passages, it will not refer to matter in the technical sense. The word appears a couple of times in Job and a few times in various books of the Apocrypha. Origen quotes the one case that probably does refer to matter in the technical sense. It is found in the apocryphal book, *The Wisdom of Solomon*. He expresses some doubts about the book, suggesting that it is not considered authoritative by everyone. That is, it was not universally accepted as part of the canon of Scripture.

Origen states that many think that Genesis 1:1–2 refers to shapeless or formless matter. After all, what else could "invisible and not arranged" refer to except formless matter? However, Origen argues that this is not matter in the technical sense. Whatever this material was, it was changeable and possessed qualities that formless matter could not.

He speaks of some ancient Greek natural philosophers who thought that everything was made of immutable, indivisible particles called atoms. This was particularly associated with the Epicurean school of philosophy. He also speaks of the material monists, who thought that a single element was the basis of all physical objects. Thales of Miletus, for example, thought that everything was made out of water. Both of these views were present in the ancient world but would lose favor with the rise of Neoplatonism.

Both of these views postulate that the underlying material of all physical objects could exist by itself. However, the material was not the formless matter of classical science but some other material that had qualities or properties. For example, if all things were made of water, then that water itself would have a form and properties. On the other hand, if the world was made of different combinations of atoms, then those atoms themselves had forms and properties.

Origen emphasizes agreement between Christians and atomism and material monism. His examples are the rejection of eternal matter and the transformation of elements from one to another. However, the atomists did not reject eternal matter. Rather, they generally believed that atoms were eternal and unchangeable. Any changes we might observe are merely the rearrangement of these atoms. Origen is thus exaggerating the agreement between Christians and atomists.

Origen insists that substance never exists without quality and thus never exists as formless matter. However, he explains that we do not know of this matter from direct experience. Rather, we can discern the existence of matter only through careful reasoning. Origen goes on to argue that the Scriptures do refer to the classical notion of matter derived from this careful reasoning. He appeals to a Psalm:

> Your eyes saw my unformed substance; in your book were
> written, every one of them, the days that were formed for me,
> when as yet there was none of them. (Psalm 139:16 ESV)

However, rather than God seeing the unformed substance of the Psalmist, Origen understands the text as referring to the Psalmist seeing God's imperfection. He argues that matter itself lacks forms and qualities and is thus imperfect; hence, it could be described as God's imperfection. The Psalmist has seen God's imperfection in the sense that he has grasped the existence of matter with his mind.

Origen also cites the Book of Enoch, in particular two quotations: "I have walked on even to imperfection," and "I beheld the whole of matter." He takes these as references to understanding the nature of matter and the difference between matter and form. While accepted by some, such as Origen, as Scripture, it would later be rejected as non-canonical by most of the Church. Today, there are only fragments of the Greek, Latin, and Aramaic versions of this text remaining. However, the Ethiopian Orthodox Tewahedo Church considers it to be inspired Scripture and has maintained manuscripts of the book in the Ge'ez language, sometimes known as Classical Ethiopic. Origen's quotations do not clearly align with anything in the extant Ge'ez text. However, it has been suggested that it is from a version of Chapter 17 where Enoch is shown the physical world from the uttermost depths to the setting place of the sun.

Origen's approach seems discordant. At first, Origen is careful to argue that Isaiah and Genesis do not speak of matter. However, then he interprets the text as referring to matter on the barest of pretexts in Psalms and the Book of Enoch. But the consistent feature is that his scientific views drive his exegesis. He rejects the idea that certain Biblical texts refer to matter because they make statements that do not align with his understanding of matter. Nevertheless, he wants to find the idea in the Biblical text and therefore reads it into these other passages.

Ephrem the Syrian

Ephrem discusses this issue in his commentary while dealing with the apparent multiple creations of the heavens and the earth during creation week (*On Creation*, Chapter 1). Genesis 1:1 indicates God created the heavens and the earth before the rest of creation. However, the heavens and the earth appear to have been created again on the second and third days of the creation week. Ephrem resolves this difficulty by arguing that God created the substance or matter that would make up the heavens and the earth before the rest of creation but did not yet give them their form. Ephrem definitively rejects resolving this difficulty by understanding the references as symbolic or allegorical.

Ephrem emphasizes that the substances of the heavens and the earth are the only things that were created at this time. According to him, even the elements did not yet exist. He bases this on Moses's silence; if the elements had been created at this time, Moses would have said so. Further, this shows that the matter making up heaven and earth must have come from nothing because there were no elements for it to come from.

Ephrem writes in Syriac, a language related to Aramaic which is similar to Hebrew. Atypically for Christians at this time, he was not limited to the Septuagint. As such, he does not think that the text indicates an invisible earth but rather cites the Hebrew words and indicates that they mean void and desolation. This void and desolate matter existed prior to the elements, suggesting something very much like formless matter.

Basil of Caesarea

Basil considers the Septuagint's description of the primitive earth: it was invisible and unfinished.[6] Basil accounts for the earth being unfinished because it lacked vegetation. He makes several attempts to explain why the earth would be invisible. Firstly, he proposes that it could not be seen because there was nobody to see it. Man was not yet created, and thus the earth was invisible in the sense that there was nobody for it to be visible to. Alternatively, he suggests that water covered the surface, hiding the solid ground below. He elaborates, pointing out that some things are innately invisible while others are only rendered invisible due to being hidden. He suggests that the earth was invisible in this second sense. Furthermore, he points out that light did not yet exist, which would have made it even harder to see the earth. As such, according to Basil, it is not surprising that the Scripture refers to the earth as invisible.

Ambrose of Milan

Ambrose discusses the initial state of the world while responding to those who claimed that matter was eternal.[7] He does not discuss the notion of formless matter but is rather concerned with addressing the objection that God should have made the world in a fully finished state. In the creation account, God first creates the world and then adorns and finishes it. But if God is omnipotent, why did he not simply create it in a fully adorned and finished state?

Ambrose gives three reasons why God would create the world in this way. First, it demonstrates that the world is not eternal by revealing that the world underwent a process of creation. Secondly, it provides an example for us to follow of

---

[6] Basil of Caesarea, "Hexaemeron," Homily 2.

[7] Ambrose of Milan, "Hexameron," The Second Homily, Chapter 7, Section 26.

performing a task step by step. Thirdly, it makes it clear that God is responsible both for the creation of the world and its adornment.

In this homily, Ambrose gives a brief explanation of the invisibility of the earth: it was covered with water, and daylight did not yet exist. However, he returns to the question in his next homily, stating that some may not have found his explanation of the invisible earth compelling.[8] Ambrose calls on those raising this issue to cease stirring up contentious disputes. It seems likely that he received some push-back on this point after his first homily.

He argues that Moses spoke from a human point of view because God was obviously able to see the earth even if it was covered with water. He points to the appearance of dry land, arguing that it only makes sense to describe the dry land as appearing if it was previously hidden. Ambrose takes this to confirm his explanation of the invisible earth. Furthermore, Ambrose points out that Moses repeats himself. Firstly, God commands that the waters be gathered together, and then the text indicates that this happened. Ambrose suggests that the repetition is to prevent disputes and emphasize that the earth was invisible due to being hidden and not that it was invisible by nature.

John Chrysostom

Chrysostom first asks why the heavens were created "bright and finished," whereas the earth was unfinished.[9] Chrysostom answers that the sky was created finished and the earth unfinished to show God's power. If God had created everything in

---

[8] Ambrose of Milan, The Third Homily, Chapter 2, Section 7.
[9] Chrysostom, *Homilies on Genesis*, Homily 2, Paragraph 12.

an unfinished state, we might think he lacked the power to do it all at once.

The earth is the provider of all things to mankind. However, this might have led to humans' treating the earth with respect beyond its due. They might treat it as an idol. Thus, God created the earth unfinished so that we would know that the gifts of the earth do not come from the earth but from the one who brought it into existence.

Chrysostom explains the invisibility of the earth by saying that it is concealed by darkness and water. According to him, Genesis 1:2 begins by stating that the earth was invisible but then explains it by speaking of the waters and darkness that covered the earth. As he points out, at that time everything was darkness and water, as even light had not yet been created.

Augustine of Hippo

Augustine addresses the subject while responding to the Manichaeans.[10] The Manichaeans asked how God could have made heaven and earth if the earth was already invisible and disorganized? Augustine criticizes them for trying to attack the Scriptures without understanding even the clearest point in Scripture. The text indicates that God first created heaven and earth, and then the earth was invisible and disorganized. Only afterwards did God bring order and organization to the world. The Manichees' objection ignores the sequence in the text.

The Manichees also asked where the waters in Genesis 1:2 came from.[11] It is true that the text nowhere explicitly says that God created the waters. Augustine suggests that if they had come to the text with an intention of piety rather than insult, then they could have understood the text. Instead, these

---

[10] Augustine, "Two Books on Genesis against the Manichees," Book 1, Chapter 3.

[11] Augustine, Book 1, Chapter 5.

critics came to the text looking for some pretext to ridicule it. Augustine argues that the waters referred to at the beginning of creation were formless matter. In order to support his interpretation, Augustine cites the passage from the Wisdom of Solomon, previously discussed in this chapter, which indicates that God created the world from formless matter.

Nevertheless, Augustine has difficulty defending this interpretation of the creation account.[12] The text speaks in the first verse of God creating the heavens and the earth. But if Augustine's interpretation is correct, God did not create the heavens and the earth at that time. Rather, he only created the formless matter from which he would later form the heavens and the earth. How does it make sense for the text to say that God created the heavens and the earth when they were only created later?

Augustine argues that it makes sense because the later creation of the heavens and the earth was already certain. A familiar example of a similar figure of speech is the phrase "you're a dead man." When someone uses this phrase, they do not literally mean that the person they are speaking to is already dead, but they may as well be dead because their death is certain. Likewise, Augustine appeals to an idiom: *iam factum puta* which means "consider it done." A request might receive the response "consider it done" or even simply "it is done" to express the idea that the request will certainly be completed.

Augustine appeals to a particular passage that he thinks uses this technique:

> No longer do I call you servants, for the servant does not know what his master is doing; but I have called you friends, for all that I have heard from my Father I have made known to you. (John 15:15 ESV)

---

[12] Augustine, Book 1, Chapter 7.

This verse speaks of Jesus' telling his disciples everything that he has heard from his Father. A simple understanding of this is that Jesus is indicating that he no longer has to keep secrets from his disciples. However, a more wooden reading would suggest that Jesus had already told his disciples absolutely everything. But this is clearly not the case, so Augustine understands it as a statement that Jesus will reveal everything in the future. That future revelation was certain but had not yet taken place.

However, such an interpretation makes little sense in the context of either passage. The purpose of this figure of speech is to emphasize the certainty of the outcome in question. But there seems to be no reason in either passage for such an emphasis.

Furthermore, the difficulties with Augustine's interpretation continue. Augustine thought that only formless matter existed at this time, but the text describes both the earth and the waters. To resolve this, Augustine argued that all references to the earth and to the waters were actually references to formless matter. He argues that it makes sense to refer to formless matter by the name of the least beautiful of the elements, earth. He thinks that the descriptions of the earth as invisible and disorganized make sense as descriptions of formless matter. Further, it also makes sense to describe the formless matter as water, either because all living things would eventually be formed and nourished by water or because water is very malleable. He further argues that the text describes formless matter in different ways to convey that it is formless matter and not ordinary elements.

Augustine returns to this subject in a later commentary. He rejects the interpretation that earth and water were created before the creation days, insisting instead that the text must be referring to formless matter. He argues that it does not make sense for these corporeal forms to be created before the beginning of the creation week. If they had, the text should have said something like "let there be earth and water," but it does not. Furthermore, if they were created before the creation days, why did God not pronounce them as good, as he did for many other acts in the creation week?

However, in this later commentary, Augustine does not believe there is a period of time between the creation of the formless matter and the formation of the heavens and the earth from that formless matter.[13] Throughout this later commentary, Augustine argues that the creation account does not relay a chronological sequence of events. Augustine argues instead that all of creation was produced instantaneously. Thus, matter and form were produced at the same time, and formless matter never actually existed.

Thomas Aquinas

Aquinas argues that matter cannot have existed without a form.[14] He also rejects the idea that all matter was created in one particular form, as held by some ancient natural philosophers. In both cases, his argument relies on an Aristotelian conception of matter and form. He concludes that matter must have been created in several distinct forms.

Aquinas argues that while previous authors appeared to contradict each other regarding whether or not matter was formless, "in reality they differ but little." He correctly points out that while Augustine speaks of formless matter in the sense of matter without any form, other authors such as Basil, Ambrose, and Chrysostom referred to the lack of beauty and completeness in the initial creation. As such, Aquinas suggests

---

[13] Augustine, *The Literal Meaning of Genesis. 1*, Book 1, Chapter 15.

[14] Aquinas, *The Summa Theologiæ of St. Thomas Aquinas*, First Part, Question 66, Article 1.

that there is agreement about the absence of formless matter. However, he does not mention other authors who did hold to formless matter. He also does not mention Augustine's earlier commentaries, where he also maintained the existence of formless matter.

Aquinas cites a scriptural passage:

> The Rock, his work is perfect, for all his ways are justice. A God of faithfulness and without iniquity, just and upright is he. (Deuteronomy 32:4 ESV)

The citation suggests that because God's work is perfect, it cannot be formless. However, the verse is clearly not about matter, nor is it clear that formless matter is imperfect in any relevant sense.

Martin Luther

Luther touches on this issue while commenting on Genesis 1:2.[15] He notes that the Hebrew words *tohuw* and *bohuw* refer to things that are empty and waste. Luther states that the earth and the waters were mixed together, and no distinctions could be made between any object and another. If the whole earth were hurled "into confusion and into one chaotic heap," this would be what Moses calls *tohuw* and *bohuw*.

Luther rejects the idea that the original material might be either mere power or akin to nothing, probably referring to an idea like formless matter. He argues that it would not make sense for Moses to call such matter "the heavens and the earth." He appeals to 2 Peter 3:5, which speaks of the earth being made out of waters and through water, arguing that the earth was made out of water, showing that the original material was water and not some sort of formless matter.

---

[15] Luther, *Luther On The Creation: A Critical and Devotional Commentary on Genesis.*

John Calvin

Calvin notes that the Hebrew words, *tohuw* and *bohuw,* refer to emptiness, confusion, and vanity.[16] Moses uses the word to describe the lack of form, ornament, and perfection. Calvin suggests that if we were to take away everything that God created after this point, we would end up with rude, unpolished, shapeless chaos. He thinks that the dark abyss was part of this confused emptiness. There was nothing solid, stable, or distinct in that mass. Only once light began to shine on the world did some sort of external appearance arise.

Calvin probably does not hold to formless matter in the sense discussed in this chapter. He does speak in terms of the state of the world being shapeless and formless. However, he does not link his interpretation to that of Plato. Nor does he explicitly describe the creation of the substance or matter of the world apart from any form. Rather, he seems to have in mind a mass of mixed elements.

Conclusions

The Christian Interaction

Some of the authors, such as Justin Martyr and Ephrem the Syrian, interpret Genesis 1 as indicating that God created formless matter. In his earlier commentaries, Augustine takes the same view; however, his view changes in his later commentaries. Even so, Augustine consistently thinks that Genesis 1 refers to formless matter. In his later commentary; he simply thinks that the Genesis account was instantaneous, and thus there was no actual time when matter was formless.

---

[16] Calvin, *Calvin's Commentary on the Bible.*

It is worth emphasizing that these authors did not simply express God's action in Platonic terms but saw the Scripture as endorsing this view. Justin Martyr went so far as to accuse Plato of plagiarizing his ideas from Moses. These three authors point to somewhat different Scriptural reasons for their positions. Justin Martyr references the Septuagint translation of Genesis 1:2. Ephrem the Syrian does not use the Septuagint but does read the Genesis 1 account as indicating the creation of formless matter. Augustine points to his Latin translation, which was based on the Septuagint as well as the Wisdom of Solomon.

However, other authors instead interpreted the text as referring to the earth's being in a disorganized or unfinished state. This was the interpretation of Tertullian, Basil of Caesarea, Ambrose, and Chrysostom. These authors all used the Septuagint and had to account for the statement that the earth was invisible. They argued that the earth was covered with waters, there was no light, making it practically invisible, and it lacked the adornment of plant and animal life, making it unfinished or unfurnished. None of these authors show any awareness that anyone interpreted the text to mean that God had created formless matter.

Origen rejected the existence of formless matter. He references the existence of many who thought that Genesis 1:2 referred to formless matter. However, he argued that this was not formless matter in the technical sense. Furthermore, he tried to argue that the Scriptures referred to underlying matter in other places.

Aquinas also rejects the existence of formless matter. He explains both Augustine's instantaneous creation and the unfurnished, hidden-earth interpretation as possible readings of the text that do not imply formless matter. His arguments against formless matter are based on the assumptions of Aristotelian metaphysics.

Luther offers an interpretation much like the unfurnished, hidden-earth interpretation of prior authors. However, because he does not use the Septuagint, he does not have to explain why the earth might be invisible. Crucially, he brings up the idea of formless matter, which he rejects on a Scriptural basis. He finds it to be inconsistent with both Genesis 1 and 2 Peter 3.

Calvin speaks of the earth as being formless. However, he does not appear to mean this in the same sense as the prior authors, who held to formless matter. Rather, his formlessness seems more like an amorphous blob of mixed materials than matter without any form at all.

Modern Science

Modern science's view of matter is very different from that of classical science. Instead of the four elements, we have the periodic table with ninety-four naturally occurring elements. In classical science, matter was continuous, whereas in modern science, matter is divided into discrete units that we call atoms. The transmutation of one element into another was thought to be common in classical science but only occurs due to nuclear reactions or radioactive decay in modern science. For classical science, the elements were the basic constituent parts of reality, but in modern science, atoms are themselves made up of subatomic particles. Furthermore, with quantum mechanics, those particles do not operate as we'd intuitively expect them to. As such, the question of whether or not formless matter could exist is now moot because matter does not have the nature postulated by classical science.

Evaluation of the Response

Some historical Christians thought that the Bible taught that God created formless matter. Under modern science, such an idea is either nonsensical or wrong. How did historical Christians come to this incorrect view? Part of the answer is that they were influenced by Platonic ideas. They already believed that the world had been formed from formless matter and thus readily understood the Bible as referring to that idea. Nevertheless, there did seem to be some Scriptural support for the idea. As discussed, the Septuagint translation of Genesis 1:2 suggests something like formless matter, and the Wisdom of Solomon explicitly states that God created from formless matter. However, those sources reflect the understanding of Greek-influenced Jewish thought and not the underlying Scriptural sources. As such, this incorrect view was caused by reading flawed science into the text and failing to go back to the original sources.

Others, such as Origen and Aquinas, argued that formless matter cannot exist and that the Genesis 1 account does not describe the creation of the world from formless matter. These authors were also influenced by the science of their day, but the science had changed. The influential science of their day rejected the possibility that formless matter existed, and as such, they sought to argue that the Bible did not teach the existence of formless matter. Origen went so far as to attempt to draw support for the classical conception of matter from irrelevant Biblical passages.

However, still others interpreted the Bible as merely stating that the earth was hidden and unfinished, not that it was formless. This reading does not strictly imply that formless matter could not exist; it does not address formless matter at all. As such, science poses no problem for this reading because it neither endorses obsolete scientific theories nor makes claims now thought to be false regarding formless matter. Obviously, the creation account is scientifically controversial in a myriad of other ways, but not with respect to this issue.

What led these authors to take this reading? Most of these authors do not raise the issue of formless matter; they show no awareness of that alternate interpretation. Luther raises the issue of formless matter but rejects it on the basis of incompatibility with the text and not scientific or philosophical concerns. As such, these authors were simply following the text and, by doing so, avoided the trap of reading flawed science into it.

In this case, we have a scientific issue that is unfamiliar to the modern Christian. But what we find is an example of the pitfalls of reading Biblical texts in light of the scientific ideas of a particular age. At different times, different ideas were read into the text. In either case, the solution is to go back to the source and follow the text to see what it actually says.

Chapter 6

Are the Heavens a Rotating Sphere?

The Scientific Question

Where does the sun go at night? Every day, the sun rises in the east, travels overhead, and sets in the west. But where does the sun go between setting and rising? How does it get from its setting in the west back to the east? The moon, likewise, follows the same path, rising in the east and setting in the west. Where does it go when we can no longer see it? The planets also follow this path, raising the same question: Where do the heavenly bodies go when we cannot see them?

For those in the northern hemisphere, everything in the heavens appears to rotate around the north star. The stars near the north star are readily seen to move in a circle around it. Heavenly bodies farther away from the north star also move in a circle around the north star, but the circle is only partially visible. Their partial circle begins when they rise in the east and ends when they set in the west. What is the cause behind this coordinated movement?

Classical science explained this movement by proposing that the heavens, where the stars were located, were a rotating sphere surrounding the earth. Furthermore, the heavenly sphere turned, rotating around the earth. Consequently, everything in the heavens, including the sun, moon, planets, and stars, moved from the east to the west due to the movement of the heavenly sphere.

This explained where the heavenly bodies went: under the earth. The sun, moon, planets, and stars were all carried under the earth by the movement of the heavens. Furthermore, that movement explained why they reappeared on the opposite side from where they disappeared. The heavens simply moved all the way under the earth and kept going back to the other side.

If each star had its own movement, it would make sense that each star would move in a somewhat different direction and at different speeds. However, it makes sense why they would all move in the same direction and at the same speed together if the stars were simply following the movement of the heavens. The entire heavens rotated around the earth in a single day, causing all of the stars to appear to move around the earth in a single day.

Another piece of evidence for this view was that different stars were visible from different places on earth. If the heavens are a sphere, then at any place on earth, approximately half of the sphere is visible. However, which half is seen will be slightly different from different places on earth. For example, as one travels south, the northernmost stars will slip out of view and new stars will appear in the south.

The Biblical Discussion

There are three sets of Bible verses that might be understood as addressing the shape of the heavens. First, a number of verses refer to the circle of the heavens. Secondly, two verses speak of the heavens as being like a tent. Thirdly, there are references to the ends of the heavens.

There are a number of verses that speak of a circle associated with the heavens. They all use the same word: חוג (*chuwg*) in either a verbal or nominal form. Furthermore, these verses are the only uses of that word:

> Thick clouds veil him, so that he does not see, and he walks on the vault (*chuwg*) of heaven. (Job 22:14 ESV)

> He has inscribed a circle (*chuwg*) on the face of the waters at the boundary between light and darkness. (Job 26:10 ESV)

> When he established the heavens, I was there; when he drew a circle (*chuwg*) on the face of the deep (Proverbs 8:27 ESV)

> It is he who sits above the circle (*chuwg*) of the earth, and its inhabitants are like grasshoppers; who stretches out the heavens like a curtain, and spreads them like a tent to dwell in; (Isaiah 40:22 ESV)

The interpretation of these verses depends on what exactly *chuwg* means.

One interpretation holds that the earth was conceived of as a flat disk. This explains Isaiah's reference to the circle of the earth. It was formed by inscribing or drawing a circle upon the face of the waters. In this reading, these verses do not speak of the shape of the heavens but do describe a flat, circular earth.

However, this interpretation runs into two difficulties. Firstly, it has difficulty incorporating Job 22:14, which refers to the circle/vault of heaven. Unlike the other verses, it cannot be plausibly understood as referring to a flat earth because it is talking about the heavens. Secondly, Isaiah implies that he who sits above the circle is at a great height, because the inhabitants of the earth are like grasshoppers to Him. A flat circle does not have any height at all.

Alternatively, this is a description of the shape of the heavens rather than of the earth. Job 22:14 is explicitly speaking of the heavens. Job 26:10 and Proverbs 8:27 both speak of inscribing or drawing a circle on the face of the waters. This most closely corresponds to the division of the waters above and waters below, on the second day of creation, which created the heavens or firmament. As such, these verses are most likely also referring to the heavens. The verse from Isaiah is the only one that refers to the circle of the earth. However, the Bible frequently refers to all of creation as the heavens and the earth. As such, by earth, Isaiah may mean all creation beneath the heavens and is thus describing the boundary between the heavens and the rest of creation.

Another verse uses a related word, which is translated as "compass":

> The carpenter stretches a line; he marks it out with a pencil. He shapes it with planes and marks it with a compass. He shapes it into the figure of a man, with the beauty of a man, to dwell in a house. (Isaiah 44:13 ESV)

A compass is a tool for drawing circles.

An argument can be made that the use of this word implies a spherical heavens. The word *chuwg* refers to circles, but it appears to refer not to a horizontal circle, separating the earth from the waters. Rather, it appears to be a vertical circle, separating the heavens from the earth. This only makes sense if the division between the heavens and the earth is a sphere. However, this requires *chuwg* to always mean a complete circle. If the word instead refers to partial circles or to a general notion of division, the text would not imply that the heavens are spherical. As such, it seems that the text simply does not give a clear indication of the shape of the heavens.

Many verses describe God as stretching out the heavens, and two compare it specifically to stretching out the heavens like a tent or curtains:

> Bless the LORD, O my soul! O LORD my God, you are very great! You are clothed with splendor and majesty, covering yourself with light as with a garment, stretching out the heavens like a tent (curtain). (Psalm 104:1–2 ESV)

> It is he who sits above the circle of the earth, and its inhabitants are like grasshoppers; who stretches out the heavens like a curtain, and spreads them like a tent to dwell in; (Isaiah 40:22 ESV)

Ancient Israelite tents consisted of curtains, cloths, or skins stretched across poles. In the event that an ancient Israelite required more room, they would expand or stretch out their tent. Within that cultural context, these verses do not suggest that the heavens are like a tent, but that the action of expanding the heavens was akin to setting up or expanding a tent.

However, the Septuagint translates the curtain as skins in Psalm 104 and a vault, in the sense of having an arched roof, in Isaiah 40. This obscures the meaning of the text, since vaults and skins are not typically stretched out. As such, some would take these verses as claiming that the heavens were in some way skin-like or shaped like a vault. Nevertheless, these passages were still clearly poetry and engaged in a metaphor. Even in the Septuagint, it is not clear that these verses indicate anything about the shape of the heavens.

The Vulgate translated Psalm 104:2 as referring to *pellem*, which means skin but was often used to refer to a tent. It translated Isaiah 40:22 as referring to *tabernaculum* which means tent. Consequently, the Vulgate avoids the issues of the Septuagint in suggesting the heavens are either like a skin or a vault.

Several verses speak of the ends of the heavens:

> For ask now of the days that are past, which were before you, since the day that God created man on the earth, and ask from one end of heaven to the other, whether such a great thing as this has ever happened or was ever heard of. (Deuteronomy 4:32 ESV)

> Its rising is from the end of the heavens, and its circuit to the end of them, and there is nothing hidden from its heat. (Psalm 19:6 ESV)

> And he will send out his angels with a loud trumpet call, and they will gather his elect from the four winds, from one end of heaven to the other. (Matthew 24:31 ESV)

If the heavens are a sphere, they do not have ends because a sphere is a continuous shape. As such, it has been argued that the Bible contradicts the idea of spherical heavens because it teaches that the heavens have ends.

Alternatively, the phrases "ends of the earth" or "ends of the heavens" may be understood as idioms referring to faraway places and not the absolute ends of the earth or heavens. Indeed, a number of uses of these phrases are clearly simply referring to faraway places:

> To the choirmaster: with stringed instruments. Of David. Hear my cry, O God, listen to my prayer; from the end of the earth I call to you when my heart is faint. Lead me to the rock that is higher than I, (Psalm 61:1–2 ESV)

> They come from a distant land, from the end of the heavens, the LORD and the weapons of his indignation, to destroy the whole land. (Isaiah 13:5 ESV)

but if you return to me and keep my commandments and do them, though your outcasts are in the uttermost parts of heaven, from there I will gather them and bring them to the place that I have chosen, to make my name dwell there. (Nehemiah 1:9 ESV)

Given these verses, it seems clear that we should understand the ends of the heavens as an idiom referring to faraway places, not a statement about the physical shape of the heavens.

Historical Christians

Theophilus of Antioch

Theophilus describes the heavens as being like a dome-shaped covering over the earth.[1] As support, he cites Isaiah in the Septuagint describing the heavens as like a vault. Theophilus refers to the pagan writers who claimed that the world was shaped like a sphere. He claims that they did not know what they were talking about. They do not even know about the creation and population of the world, which can be learned from the Scriptures.

Athenagoras the Athenian

Athenagoras mentions this issue while making an argument against polytheism.[2] He assumes that the world and the heavens are spherical and makes an argument on that basis. If the world is a sphere, there are two places: within the spherical world or outside of it. If God is the creator of the world, what room can there be for any other gods? Another god cannot be in the world because it already belongs to God, its creator. But

---

[1] Theophilus of Antioch, "To Autolycus," Book II, Chapter 13.
[2] Athenagoras, "A Plea for the Christians," Chapter 6.

109

that god also cannot be outside of the world because that is already occupied by God the creator. Where else could a god be?

Arnobius of Sicca

Arnobius uses the spherical nature of the world in an argument against polytheism.[3] He speaks of deities presiding over the left and the right. He is probably referring to the practice of augurs in the ancient world. Augurs would attempt to predict the future by observing whether certain events, such as birds flying by, happened on their left or on their right. Events on the left were bad omens, whereas events on the right were good omens.

"Sinister" is the Latin word for left. The negative meaning of sinister today is at least partially a consequence of this practice. Arnobius argues that it is silly to think that whether events happened on your left or right would have any meaning. In particular, it is silly to suggest that evil gods would control the events on your left while favorable gods would control the events on your right.

He rests this argument on the world being a sphere. The world thus has neither a beginning nor an end, and every point is very much the same. Consequently, there is no left, right, east, or west part of the earth. Such names only exist in relation to the person using them. As such, it makes no sense to suggest that some god might control one part but not the other.

Lactantius

Lactantius argues for the impossibility of humans gaining certain knowledge of the natural world.[4] He compares it to

---

[3] Arnobius, "Against the Heathen," Book IV, Chapter V.
[4] Lactantius, "The Divine Institutes," Book III, Chapter III.

attempting to speculate about a distant city based only on its name. He lists a number of examples of things that he thinks cannot be known. In particular, he mentions both the idea that the stars are fixed to the heavens and whether the heavens are at rest or moving with incredible swiftness. He neither accepts nor rejects the theory of the moving heavens but insists that we cannot know whether it is true.

Later, Lactantius argues against people living on the other side of the earth, the Antipodes.[5] As part of his discussion, he explains the reasoning behind inferring a rotating heavens. He explains that the philosophers observed the setting of the sun, moon, and stars all towards the same quarter (the west). He states that they do not understand the contrivance regulating their course. We will see in a later section that Lactantius conceives of something machine-like regulating the course of the stars. Thus, they came up with the idea that the heavens themselves are a sphere turning around the earth.

Lactantius mentions the carving of an orb representing the figure of the world. A surviving example is the Farnese Atlas, a marble statue of Atlas, a Titan from Greek mythology, holding up world. He is not, as you might assume, holding up the planet Earth, but the sphere consisting of the whole world, including the heavens. The constellations are visible, carved on the outside of the sphere, because the stars were thought to be the farthest points of heaven.

Basil of Caesarea

Basil speaks of astronomers who study the stars, learning much about them.[6] They have learned the distances to the stars, the paths of the stars, and the time of the stars' return. In fact,

---

[5] Lactantius, Book III, Chapter 24.
[6] Basil of Caesarea, "Hexaemeron," Homily I.

Figure 6.1: The Farnese Atlas

these men would seem to know everything, except that which is most important, that God is the Creator of the universe. Basil mentions the stars of the southern pole. Just as all stars in the northern hemisphere rotate around the northern pole, the stars in the southern hemisphere appear to rotate around the southern pole. By mentioning this pole, Basil implies that he accepts the theory of a spherical heavens.

John Chrysostom

John Chrysostom addresses this issue while commenting on Hebrews:[7]

> Now the point in what we are saying is this: we have such a high priest, one who is seated at the right hand of the throne

---

[7] Chrysostom, "Homilies on Hebrews," Homily 14.

of the Majesty in heaven, a minister in the holy places, in the true tent that the Lord set up, not man. (Hebrews 8:1–2 ESV)

Chrysostom declares that this verse contradicts both the idea that the heavens are spherical and the idea that the heavens turn. Unfortunately, Chrysostom is not very explicit in his argument.

Many modern readers may assume that this verse is using "heaven" to refer to the realm of God and not the location of stars. Historically, the word "heaven" referred primarily to the skies above rather than the realm of God, thus making it not obviously flawed to take this as a reference to the physical heavens.

The verse indicates that Jesus is seated in *heaven* in the true *tent*. Chrysostom is probably taking this as a reference to the passages that speak of God stretching out the heavens like a tent. As such, he interprets the verse as stating that the heavens are like a tent on the earth and not a sphere surrounding the earth.

The Greek text says that God ἔπηξεν (epēxen) the tent, which is the heavens. This word means to fix, fasten, or make firm. Arguably, something fixed or fastened in place does not move, and thus Chrysostom probably takes this to imply that the heavens are not turning.

Although Chrysostom is not explicit, these are plausibly the arguments that Chrysostom intends to use to argue that the heavens are not a sphere and do not rotate. However, both arguments take the heavens as a tent, a metaphor further than can be justified by the text.

Augustine of Hippo

Augustine indicates that it was commonly asked what the

Scripture taught about the shape of the heavens.[8] He indicates that many scholars engaged in lengthy discussions on this subject. But he thinks such a discussion is not profitable. It is not important whether the heavens are spherical and surround the earth or only cover it like a dome or disk on one side. The sacred writers, in their wisdom, simply omit this subject from discussion.

However, despite declaring the subject not worth our time, Augustine engages it because the credibility of Scripture is at stake. He is concerned that someone will misunderstand the text and think that it contradicts their knowledge of the world. Consequently, that person will refuse to believe the Bible on other subjects.

Augustine claims that the sacred writers *knew* the truth about the shape of the heavens. Nevertheless, the Holy Spirit did not inspire them to write down this information, since it would be of no avail for salvation. This idea is foreign to modern Christians, who take it for granted that the Biblical authors were scientifically ignorant. However, Augustine and other historical Christians did not make the same assumption.

Augustine points out that the Scriptures compare heaven to both a skin and a vault. Recall that the Hebrew text in both cases describes the earth in terms of a tent, but Augustine is unaware of this because he could not read Hebrew. He is concerned to reconcile the Biblical text with the scientifically understood shape of the heavens. However, he does note that it is a man-made theory and might be incorrect. Truth is found in what God reveals over and against the theories of man. However, if the heavens were proved to be spherical with proofs that cannot be denied, we must show the Scripture is not opposed to such an established theory.

---

[8] Augustine, *The Literal Meaning of Genesis. 1*, Book II, Chapter 9.

Nevertheless, the more important issue is resolving the apparent conflict between describing the heavens as a vault and a skin. After all, a skin stretched out flat and a curved vault have very different shapes. We must interpret the Scriptures in such a way as to reconcile these passages.

Augustine interprets the reference to the heavens as a vault as speaking only of that portion of the heavens that we can see. Since it is only possible to see about half of the heavenly sphere at any point, we can effectively see a vault. A sphere is essentially a vault that goes all the way around. As such, this passage is compatible either with the heavens as a vault or a sphere.

Augustine has more difficulty with the passage describing the heavens as being like a skin. He briefly mentions an allegorical interpretation of this passage that he gave elsewhere. In that allegory, the skin is the Scriptures, which have been spread out over all mankind by the work of God's ministers. However, historically, allegorical interpretation was a supplement to, rather than a replacement of, a literal interpretation. Thus, Augustine feels the need to also explain the literal meaning. He suggests that this meaning should be obvious to everyone. Nevertheless, he needs to explain it because some doggedly literal-minded interpreters will miss it.

Augustine suggests that a vault could be either curved or flat. Likewise, a skin could be stretched out flat or into a spherical shape. He particularly points to the examples of a leather bottle or an inflated ball, both of which are made of skins but have different shapes.

Augustine continues by introducing the question of whether the heavens are in motion.[9] If the heavens are a firmament, how can they be in motion? However, if the heavens are

---

[9] Augustine, Book 2, Chapter 10.

stationary, how do the heavenly bodies make their path from east to west? How do the stars near the north star make their smaller orbits around that pole? The stars look like the heavens are turning either like a sphere or a disk.

The idea that the heavens are a firmament derives from the Genesis 1 account of the creation of the heavens:

> And God said, "Let there be an expanse (firmament) in the midst of the waters, and let it separate the waters from the waters." And God made the expanse (firmament) and separated the waters that were under the expanse (firmament) from the waters that were above the expanse (firmament). And it was so. And God called the expanse (firmament) Heaven. And there was evening and there was morning, the second day. (Genesis 1:6–8 ESV)

The ESV translates the Hebrew as "expanse," whereas some other translations use "firmament." We will consider the question of the meaning of the Hebrew in more detail in a later section. For now, it suffices to note that the Septuagint uses a word that conveys the same meaning as "firmament," suggesting something firm or strong.

Augustine suggests that while some have spent much effort on this question, neither he nor those he wishes to teach have the time to spend on these questions. This is similar to the comments he made regarding the shape of the heavens. Augustine argues that the Scripture's use of the word "firmament" does not demand a stationary heaven. We may understand the word to suggest not a motionless heaven but one that is solid and maintains the barrier between the waters above and the waters below. Augustine also argues that if the heavens were discovered to be immovable, the motion of the stars could still be explained. Those scholars who had carefully studied the question had concluded that all the observed features of the sky could be explained by the stars moving independently but in concert.

Severian of Gabala

Severian rejects the spherical heaven as unbiblical, describing its proponents as idle talkers.[10] He rejects both the idea that it is a sphere and that it turns on its axis. He refers to Isaiah 40:22, but instead of saying that God stretched out the heavens like a vault, he makes it even more explicit, saying that God *arched* the heavens like a vault.

Severian insists that the sun does not climb. He probably does not mean that the sun does not rise at all, but merely that it does not rise from under the earth. He thinks it goes up and down over the course of the day but does not go so far down as to go beneath the earth.

His first reason for this claim is that the "Biblical author" says that heaven has a beginning and an end. Later, he quotes Psalm 19 regarding the sun:

> Its rising is from the end of the heavens, and its circuit to the end of them, and there is nothing hidden from its heat. (Psalm 19:6 ESV)

He also quotes Jesus speaking of the ends of heaven.

> And he will send out his angels with a loud trumpet call, and they will gather his elect from the four winds, from one end of heaven to the other. (Matthew 24:31 ESV)

A sphere does not have ends. Consequently, according to Severian, the heavens cannot be spherical, and the sun does not travel under the earth.

Secondly, Severian argues that two Biblical passages speak of the sun emerging and not ascending. There are in fact many

---

[10] Severian, "Homilies on Creation and Fall," Homily III.

verses that speak of the rising of the sun; however, Severian selectively focuses on these two passages, which do not use the typical word for the rising of the sun. Instead, these verses literally refer to the coming forth or going out of the sun. This is true both in the original Hebrew and in the Septuagint translation. This same word is used to refer to everything from the growth of plants to flowing rivers and leaving a city. All are said to "come forth" despite being quite different. As such, drawing any conclusion about the movement of the sun from the use of this word is unsupportable.

Where does the sun go at night if it does not go under the earth? Since the heavens are a tent or vaulted roof on top of the earth, the sun must remain on top of the earth. Severian argues that the sun, after it sets, heads north, concealed by the waters (probably the waters above from Genesis 1) and heads, hidden, to the east so that it can rise again.

Severian appeals to Ecclesiastes in defense of his claim that the sun returns from west to east by a northern passage.

> The sun rises, and the sun goes down, and hastens to the place where it rises. The wind blows to the south and goes around to the north; around and around goes the wind, and on its circuits the wind returns. (Ecclesiastes 1:5–6 ESV)

However, Severian quotes the passage as saying:

> The sun rises, and the sun sets; on rising, it travels to its setting, and goes around to the north. It goes round and round, and rises in its place.

Severian is simply misquoting this verse. The verse speaks of the sun rising and setting. It also speaks of the wind blowing north and south. Severian conflates the two to produce a claim that is not in the text, even in the Septuagint.

Severian seeks to argue from Scripture that the heavens are not spherical. However, he relies on some faulty translations in the Septuagint. Furthermore, he misquotes the text, giving readings that support his views. He selectively quotes verses that appear to support his view. He assumes that ends must be literal and not idiomatic. Severian's exegetical methods leave much to be desired.

Cosmas Indicopleustes

Cosmas leaves no doubt as to his opinion. The heavens are not spherical, and only fake Christians think otherwise.[11] According to him, they look down upon Scripture and despise it.

Cosmas appeals to the movement of the planets.[12] The movement of the stars was well explained by understanding the heavens as a rotating sphere. However, the planets did not follow this same movement. They did move around the earth along with the other stars, but they also traced a circle in the opposite direction. Cosmas asks how this can be if the stars and planets are all made of the same material, aether. Furthermore, since aether is a simple substance, that is, not made of a combination of other elements, different parts should not have different properties.

Cosmas presses further, asking why these planets sometimes slow down and reverse direction. If they, by nature, have a movement opposite to the rest of the heavens, why do they sometimes instead stand still or move in the opposite direction? Since the heavens are entirely made of aether, nothing can impede their progress. What force could be present that forces them to move against their nature?

Cosmas rejects the idea that this might be an optical illusion. A small movement might be excused as an illusion or a mistaken observation. However, as Cosmas points out, the

---

[11] Indicopleustes, *Christian Topography*, n.d., Prologue II.

[12] Indicopleustes, Book I.

planets are observed to move backward far enough to shift their way along the zodiac. The zodiac is the set of constellations lying along the paths of the sun, moon, and planets.

The explanation given for the backward movements of the planets in classical science was that they were moving on epicycles. The planets did not revolve around Earth but around a point that revolved around Earth. The famous astronomer Claudius Ptolemy had worked out a system of such epicycles that accurately, for the time, predicted the positions of the planets.

Cosmas objects that it makes little sense for the planets to be carried along on epicycles or vehicles. Surely, they are capable of movement on their own; they do not need to orbit around other points. Cosmas finds the idea ridiculous. Cosmas further asks why the sun and the moon do not have epicycles. Are they not worthy of epicycles? Did the Creator run out of material? Clearly, this cannot be the case.

Cosmas points to real weaknesses in the cosmology of classical science. Indeed, the issue of explaining the movements of the planets, including those retrogressions and pauses, would ultimately lead to the rejection of geocentrism and the adoption of heliocentrism.

Cosmas invokes Isaiah's statement in the Septuagint that the heavens were like a vault to show that the heavens were not spherical. However, he also quotes Job as saying:[13]

> He has inclined heaven to earth, and it has been poured out as the dust of the earth. I have welded it as a square block of stone

That passage is very different in a modern English translation:

---

[13] Indicopleustes, *Christian Topography*, n.d., Book II.

Who can number the clouds by wisdom? Or who can tilt the
waterskins of the heavens, when the dust runs into a mass
and the clods stick fast together? (Job 38:37–38 ESV)

There is some question about the meaning of these verses. The
most common interpretation is that the verse is pointing out
that Job cannot make it rain. The second verse indicates either
parched ground before it rains or the effect of rain on the
ground. In either case, it does not appear to be a reference to
the heavens being glued to the earth.

It is difficult to discern what the Septuagint means. It
speaks first of the heavens being inclined to earth. Then the
heavens are poured out like the dust of the earth. Finally, they
are welded like stone. Cosmas takes this as describing the
heavens being placed on and cemented to the earth. However,
his reading does not really make sense of the passage. In particular, it leaves the reference to pouring out like dust entirely
mysterious.

Cosmas asks, if the heavens are a rotating sphere, what
happened to Jesus as he ascended into heaven?[14] He mocks
Christians who agreed with the rotating sphere, suggesting
that they actually worship the moon and believe that Jesus
ascended to be with their goddess. He asks whether Jesus
cuts through the solid heavens, moving opposite to them, or
is violently whirled around with the direction of the heavens.
Both are obviously nonsense. This argument assumes that
Jesus ascended into the physical heavens rather than the spiritual abode of God. Like Chrysostom, he does not distinguish
between Heaven and the heavens.

Cosmas invokes Isaiah 40:22 which, as discussed previously, compares the heavens to a vault in the Septuagint. He
also cites Isaiah as saying, "The Lord God who made the

---

[14] Indicopleustes, *Christian Topography*, n.d., Book VII.

heaven and fixed it" which establishes that the heavens are not in revolution. This appears to be from:

> For thus says the LORD, who created the heavens (he is God!), who formed the earth and made it (he established it; he did not create it empty, he formed it to be inhabited!) "I am the LORD, and there is no other. (Isaiah 45:18 ESV)

However, this passage does not speak of the heavens being fixed either in the Hebrew or the Septuagint.

Cosmas wrote a long and repetitive book defending his cosmology. Here, we have considered a few passages. We must grant that he makes some valid points in his scientific argument against the spherical heavens. Nevertheless, he fails to address the arguments for a spherical heavens. His Scriptural arguments depend on flawed translations and dubious exegesis.

Conclusion

The Christian Interaction

A number of early authors simply assumed that the earth was a sphere. This includes Athenagoras the Athenian, Arnobius of Sicca, and Basil of Caesarea. They invoked the idea of a spherical heavens in their arguments. They did not bring up any Scriptural or theological objections to the spherical heavens. Rather, they viewed the spherical heavens as an established fact of science and showed no signs of awareness that anyone might view them as in conflict with the Scriptures.

On the other hand, some rejected the idea that the earth was a sphere on Biblical grounds. This included Theophilus of Antioch, Severian of Gabala, John Chrysostom, and Cosmas Indicopleustes. The two common verses invoked to argue that the heavens were not a sphere were Psalm 104:2 and Isaiah

40:22. Both speak in the Hebrew text of the heavens being stretched out like a tent. However, the Septuagint states that the earth is stretched out like a skin or a vault. Severian and Cosmas both argue that references to the ends of the earth or heavens show that the heavens are not spherical.

Augustine of Hippo engages the Scriptural arguments against the sphere. He attempts to reconcile the idea of the heavens as a skin or vault, both with each other and with the theory of the spherical heavens. Likewise, he argued that the heavens being a firmament is compatible with the heavens being in motion.

Most later authors do not discuss the question of whether or not the heavens are spherical. Rather, they simply accept, without question, that it is spherical. They show no awareness that there could be or had been a Scriptural objection to the idea.

Modern Science

According to modern science, rather than a rotating spherical heavens, we have a rotating spherical earth. To some degree, this is simply a matter of perspective. The boundary between the planet Earth, including its atmosphere, and outer space is spherical. If viewed from the perspective of the earth, the heavens are a sphere enclosing the earth. On the other hand, if viewed from the perspective of outer space, the earth is a sphere inside space. As such, whether the earth is a sphere or the heavens are a sphere is simply a matter of perspective.

Classical science thought that the heavens rotated around the earth, but modern science holds instead that the earth rotates. However, this is also a matter of perspective, according to modern science. In modern science, movement, including rotation, is never absolute but always relative to something else. Thus, it is accurate to describe the earth as rotating relative to the universe or the universe as rotating relative to the earth.

Classical science thought that the universe as a whole was a sphere. Modern science regards the universe as a vast expanse that we cannot see the edges of. Many argue that the universe is, in fact, infinite. Whatever the shape of the universe, it is clear that it is not the sphere envisioned by classical science.

Evaluation of the Response

There were Christians who thought that the theory of a rotating spherical heavens contradicted Biblical teaching. The theory of the spherical heavens was not entirely correct. However, these objections centered on the two key points, which have been confirmed by modern science after adjusting for perspective. From the perspective of someone on earth, the heavenly bodies do go under the earth, and the heavens are in motion. As such, these Christians have not been vindicated by modern science.

What led to this incorrect conclusion? Part of the issue was the dependence on the Septuagint, which translated key words in a way that suggested the heavens were shaped like a skin or a vault. However, this is not the entire reason, because many authors depending on the Septuagint nevertheless did not think it was making a statement about the shape of the heavens. Their conclusions were rooted in poor exegesis. The most extreme example is Severian, who simply misquotes the text. They took references to the heavens being pitched like a tent as a description of the shape of the heavens when it is quite clearly a metaphor that doesn't make a statement about the shape of the heavens. They understood the phrase "ends of the earth" as referring to the absolute ends of the earth rather than the relative ends of the earth, as is clear from other usages in Scripture.

Most of the earlier and later authors accepted the spherical heavens without question. They show no awareness that anyone might have a Scriptural objection to that sphere. However, in the fourth and sixth centuries, we have authors engaged in dubious exegesis in an attempt to show the heavens are not a rotating sphere. Other Christians responded, and the idea died out. Ultimately, the Church did not find the case compelling and rejected it.

The idea of a rotating heavens is unfamiliar to the modern Christian. However, many will be familiar with accusations that the Church historically defended a flat-earth cosmology. There were indeed Christians who rejected classical cosmology. However, this rejection is rare, and we see that it results from failing to go back to the source. When we return to the Hebrew text and compare how similar language is used throughout the Bible, it is clear that they did not have a case. Indeed, the Church came to the same conclusion, accepting classical cosmology.

Chapter 7

Where is the Bottom of the Earth?

The Scientific Question

If the heavens are spherical, does this imply that the earth is also spherical? Strictly speaking, it is conceivable that the heavens could be a sphere, while the earth itself could be a flat disk (or various other shapes) contained within that sphere. However, it made sense that the earth would be a sphere like the heavens that enclosed it. But if the earth is a sphere, why don't the people living on the sides or bottom of it fall off? If we conceive of the earth as a giant ball, it makes intuitive sense that people could live on the top of that ball, but it would seem that people would fall off the sides or the bottom.

Classical science proposed that the bottom of the earth was actually the center of the earth. The entire universe was a sphere, and all heavy objects moved, by their nature, towards the center of that sphere. The direction that we think of as down is actually towards the center of the universe.

The earth was very heavy, and thus, wherever it began, it had long since fallen to the bottom or center of the universe. Likewise, all heavy objects had long since fallen onto the earth. The solid mass of the earth rested because it was equally distributed on all sides of the center and thus pushed equally on all sides, keeping the earth immobile and pushing it into a roughly spherical shape.

Consequently, people standing on the opposite side of the earth were not actually on the bottom of the earth. Rather, the bottom of the earth was in the center of the earth. People standing on the other side of the earth were also on top of the earth. All of the places on the surface of the earth's sphere were at the top of the earth, and it was possible to stand anywhere on the surface (assuming there was land to stand on).

Furthermore, it seemed natural to assume that the other side of the earth would be much like this side. It ought to have land, oceans, mountains, forests, animals, and even other people. The ancient Greeks called this land and its inhabitants the "Antipodes," referring to their feet being opposite to the Greek's feet.

The Biblical Discussion

Nothing in the Scriptures explicitly addresses the nature of gravity. Nor are there any passages that imply either acceptance or rejection of classical gravity. However, there are statements that speak of the heavens' being above the earth.

> You shall not make for yourself a carved image, or any likeness of anything that is in heaven above, or that is in the earth beneath, or that is in the water under the earth. (Exodus 20:4 ESV)

> And as soon as we heard it, our hearts melted, and there was no spirit left in any man because of you, for the LORD your God, he is God in the heavens above and on the earth beneath. (Joshua 2:11 ESV)

> For as high as the heavens are above the earth, so great is his steadfast love toward those who fear him; (Psalm 103:11 ESV)

But is it accurate to describe the heavens as being above the earth? If the heavens are a sphere, are they not rather around the earth? It would seem that the heavens are both above and below the earth, not merely above it.

However, according to the classical science conception, the bottom of the universe is the center of the universe. Down is towards that center. Up is away from that center. In contrast, the other side of the earth is not the bottom of the earth. Consequently, the heavens are still above the earth, because the heavens all around the earth are further away from the center of the universe than the earth is.

The same basic argument would apply to variations on this theme. For example, we could consider verses that speak of birds flying above the earth. Birds do, in fact, fly above the earth, so long as we understand the bottom of the earth to be the center of the earth. All such similar cases proceed along the same line of argument.

Historical Christians

Lactantius

Lactantius thinks it is senseless to suggest that people might be living on the other side of the earth.[1] They would be upside down. Their feet would be above their heads. Things lying down would actually be hanging from the bottom of the earth. Trees and crops would grow downward. Rain, snow, and hail would fall upward onto the earth. He mentions the hanging

---

[1] Lactantius, "The Divine Institutes," Book III, Chapter 24.

gardens, one of the seven wonders of the ancient world, asking why hanging gardens would be at all impressive if there are hanging mountains, fields, seas, and cities on the other side of the world.

Lactantius makes a valid logical point. When an assumption leads to absurd or incorrect conclusions, that assumption must be false. On the other hand, when someone is convinced of an incorrect assumption, they are often forced into defending absurd or implausible positions. They ought to reevaluate their original position. He thinks that the conclusion that there are people on the bottom of the earth is absurd, and thus those coming to conclusions should have realized that their starting assumptions were incorrect.

Lactantius goes on to explain the argument for the existence of the Antipodes. We previously considered his explanation of the argument for the spherical heavens. If the heavens are round and enclose the earth, it follows that the earth must also be a globe. However, if the earth is round, it makes sense that it would be similar on all sides. There ought to be mountains, plains, and seas in all parts of the earth. Furthermore, there must be men and other animals on the other side of the earth.

But why does not everything fall off the bottom of the earth? Lactantius explains the classical science theory of gravity. The nature of things is that heavy bodies fall towards the center of the universe, whereas light bodies move away from the center. Lactantius finds this explanation absurd. He wonders if the people who suggest this idea are deliberately defending falsehoods for the fun of it. He claims that he could present many arguments that show it is impossible for the heavens to be lower than the earth. Unfortunately, he does not go into any detail as to what exactly these arguments would be.

Lactantius's arguments come down to finding it absurd that people would be upside down on the other side of the earth or that the heavens would be below the earth. However, if the bottom of the earth is the center of the earth, neither of these is true. The heavens are still above the earth, and the people living on the other side are still right side up.

Notably, Lactantius invoked neither Scripture nor theology in his argument. Indeed, in context, his argument was that philosophers believed in nonsense, of which the Antipodes are but one example. His concern was not whether or not the belief in the Antipodes was compatible with Christian belief. Furthermore, Church Fathers such as Jerome found that he lacked accurate knowledge of Christian belief:[2]

> Lactantius has a flow of eloquence worthy of Tully:[3] would that he had been as ready to teach our doctrines as he was to pull down those of others!

Basil of Caesarea

Basil addresses this issue while discussing the question of what supports the earth.[4] In context, he is critiquing the ability of the philosophers of his day to account for the workings of the world. He describes the classical account, that the earth is kept still in the center of the universe, resting on itself rather than any external support. Basil dismissively describes this theory as a "great display of words."

Basil explains that, according to this theory, the earth is not arbitrarily at the center but instead is there because of the tendency of all heavy things to move to the center of the

---

[2] Jerome, "Letter 58."

[3] Tully is a reference to Marcus Tullius Cicero, most commonly referred to by modern authors as Cicero.

[4] Basil of Caesarea, "Hexaemeron," Homily I, Section 10.

universe. All heavy objects, regardless of where they may have started, have fallen to the center of the universe. It is thus not surprising that the earth rests at the center of the universe. It rests in its natural place at the bottom or center of the universe.

Despite having described this system as a "great display of words," Basil appears to find this theory acceptable. He says that if his listeners find the account probable, they should admire God for designing such a wonderful mechanism. The wonder of a stable earth is not lessened because we can scientifically explain why the earth is stable.

Basil neither endorses nor rejects the classical theory of gravity. He calls it a great display of words but allows that it might seem probable to some of his listeners. Certainly, he brings up no Scriptural or theological objection to it.

Augustine of Hippo

Augustine addresses an objection to the Christian faith, arguing that bodies cannot be present in heaven, because they would fall down.[5] He is not writing of the departed Christians, who currently lack bodies. Rather, he is speaking of the resurrected body of Jesus Christ and, after the Second Coming, the bodies of all the saved. He believes that in the world after the Second Coming, Christians will live not on the new earth but in the new heavens. This leads to the objection that bodies must, by their nature, fall to the earth and could not remain in the heavens.

Augustine first points out that the earth is unsupported. How can it be objected that physical bodies cannot remain unsupported in heaven if the earth itself is unsupported? However, Augustine acknowledges that the notion that all heavy bodies gravitate toward the center of the universe might

---

[5] Augustine, "The City of God," Book XIII, Chapter 18.

provide an answer to his objection. This is a description of the classical understanding of gravity. Furthermore, the tendency of all things to fall towards the center extended throughout the whole universe. Everything, no matter how far away from the earth, fell towards the earth, and this would presumably include the bodies of the resurrected saints. As such, there is an apparent conflict between the classical conception of gravity and resurrected physical bodies residing in the heavens.

Augustine argues for the possibility of God creating bodies that do not fall to the earth. In particular, he points to Plato's thinking that a lesser god could remove the burning from fire. The idea was that the light of the eyes was a fire, but not one that burned. If this were possible, God could surely make bodies that would not fall to earth. Additionally, if God could bring together the body and soul, surely he could also remove the weight from the body.

Augustine thinks that the existence of people living in the Antipodes is not credible.[6] The ancient world did not have a historical record of anyone's visiting the Antipodes and returning. Their existence was purely conjectural. Further, Augustine shows that the argument for the Antipodes does not follow. The earth being spherical does not mean that there is land on the other side. Even if there is land on the other side, this does not mean that there are people there.

Augustine argues that there cannot be people living on the other side of the world because there is no way for people to have reached it. As such, any inhabitants of the Antipodes cannot have descended from the first man, Adam. He insists that the Scripture proves its historical statements by its successful prophetic predictions. It would, Augustine believes, be absurd to think that anyone crossed the ocean to reach the Antipodes.

---

[6] Augustine, Book XVI, Chapter 9.

Augustine has no problem with the physics of people standing on the other side of the earth without falling off; he simply thinks that nobody could have reached it.

In both cases, Augustine does not endorse the classical theory of gravity. But he accepts it for the sake of argument. He does not appear to be aware of any Scriptural objections to the idea.

Cosmas Indicopleustes

Cosmas mocks those who believe in the Antipodes.[7] They believe in upside-down people living on the underside of the earth. He presents this as a silly attempt to explain the purpose of the sun going under the earth. As Cosmas sees it, rather than accept the truth of the Scriptures, these people have chosen to believe in something obviously absurd.

Cosmas mocks the idea of the Antipodes, suggesting they are old wives' fables. If two people stood on opposite sides of a sphere, they could not both be upright. One must be right side up while the other is upside down. When it rains, rain would have to fall down on one and up on the other. But this, Cosmas insists, is ludicrous.

Cosmas alludes to the book of Acts:

> And he made from one man every nation of mankind to live on all the face of the earth, having determined allotted periods and the boundaries of their dwelling place, (Acts 17:26 ESV)

He argues that there are people on the singular face of the earth. In Cosmas's thinking, the people on the bottom of the earth live on a separate face from those living on the top of the earth. But the Bible indicates that all people live on one face.

Cosmas refers to Philippians:

---

[7] Indicopleustes, *Christian Topography*, Book I.

> so that at the name of Jesus every knee should bow, in heaven
> and on earth and under the earth, (Philippians 2:10 ESV)

He argues that "those above the earth" refers to angels, "those on the earth" refers to living men, and "those below the earth" refers to those who are buried in the earth. For Cosmas, the implication is that there cannot be people living in the Antipodes because they are neither living above the earth nor dead and lying below the earth.

Cosmas refers to the Gospels:

> Behold, I have given you authority to tread on serpents and
> scorpions, and over all the power of the enemy, and nothing
> shall hurt you. (Luke 10:19 ESV)

Christ gave his followers the power to tread on serpents and scorpions. But to tread upon anything, whether the earth, or scorpions, only makes sense if the one treading is above those things. However, if the notion of above and below is no longer applicable in a spherical world, this no longer makes any sense.

Cosmas illustrates his argument with a drawing depicting four men standing on four sides of the earth. He argues that the four men cannot all be standing upright. Only one of them can be standing upright. It is silly, therefore, to claim that they are all standing upright.

Cosmas's argument does not progress beyond declaring the classical conception to be absurd. He invokes various scriptural passages, but in every case, he simply reiterates the same basic assertion. Nothing in these passages is actually incompatible with the classical science conception of down as towards the center of the earth and up as away from the center. Cosmas is reading his own rejection of classical gravity into the text.

Martin Luther

Luther describes it as "wonderful" that the philosophers of his day had determined the earth to be the center of the universe.[8] It cannot fall because it is supported by all of the other spheres surrounding it. All of the other spheres and heavens rest on the earth. Nevertheless, Luther insists that this stability rests ultimately on the Word of God.

John Calvin

While discussing what keeps the earth in place, Calvin writes:[9]

> This I indeed grant may be explained on natural principles; for the earth, as it occupies the lowest place, being the center of the world, naturally settles down there. But even in this contrivance there shines forth the wonderful power of God.

Calvin allows that the classical science theory may be correct and that the center of the earth may be the center and lowest place in the world. However, he insists that even in this, it shows the wonderful power of God.

Conclusions

The Christian Interaction

Lactantius and Cosmas both reject classical gravity. Both do so because they think that the idea is clearly absurd or ridiculous. Cosmas attempts to invoke many Biblical passages, but in every case, his conclusion still comes down to finding the classical conception of gravity too counter-intuitive to accept.

---

[8] Luther, *Luther On The Creation: A Critical and Devotional Commentry on Genesis*, 1:Chapter 1.

[9] Calvin, *Calvin's Commentary on the Bible*, Commentary on Psalm 104.

Basil, Calvin, and Augustine neither accept nor reject the theory. Luther appears to accept it. Basil disparages the idea somewhat but regards it ultimately as acceptable. Augustine accepts it for the sake of argument. None has any Scriptural or theological objections to it.

Modern Science

Classical gravity was incompatible with heliocentrism. If the sun were the center of the world, then all things ought to fall towards the sun and not towards the earth. Either the earth was the center of the universe or the classical conception of gravity was incorrect. The earliest heliocentrists did not have an answer to this difficulty, but Johannes Kepler would argue for the basic idea: heavy objects did not fall towards the center of the universe; they were attracted to other heavy objects. Humans fall to the earth because the earth is very heavy. This idea would be refined, but the same basic idea is held to be true by modern science.

Classical gravity is thus, strictly speaking, incorrect. There is not a single center of the universe towards which all heavy objects fall. Instead, objects fall towards nearby massive objects. Nevertheless, classical gravity is approximately correct for people living on earth. While the gravity of the sun and the moon do affect the earth, it is the earth's gravity that plays the most significant role in our daily lives.

Evaluation of the Response

The acceptance of classical gravity for the sake of argument has fared well. Insofar as modern science agrees with classical gravity, these arguments still work. Basil's disparagement of the theory of classical gravity has not aged well. Nevertheless, since he still finds the idea acceptable, he is not entirely wrong.

Lactantius and Cosmas, on the other hand, were both wrong in their rejection of classical gravity. In both cases, this seems to have been rooted in finding the idea absurd. Only Cosmas attempted to make a Scriptural argument. However, even there, his arguments are still effectively that he finds the idea of classical gravity ridiculous.

It is also worth noting that neither Lactantius nor Cosmas is a major figure. Both are on the fringes of Christian belief. This is as opposed to Basil and Augustine, who were influential Church Fathers. The rejection of classical gravity seems to have been a fringe position within Church history. However, we know that some others, such as Chrysostom, who are well respected Church Fathers, did adhere to some aspects of what we would today describe as a flat-earth cosmology, and likely would have also rejected classical gravity.

We see in this issue a case where there was some debate, but the Church concluded that the Bible was compatible with classical science. The Bible simply did not teach anything about the nature of gravity and, as such, was compatible with different conceptions of the nature of up and down. When we examine the sources of those who rejected classical gravity, we find it not to be from Scripture but instead from the authors' own rejection of the idea as obviously absurd.

Chapter 8

How Many Heavens Are There?

The Scientific Question

According to a common legend, Alexander the Great wept because there were no more worlds to conquer. Curiously, this is close to the opposite of what the ancient texts actually record. Plutarch wrote:[1]

> It is reported that King Alexander the Great, hearing Anaxarchus the philosopher discoursing and maintaining this position: That there were worlds innumerable: fell a-weeping: and when his friends and familiars about him asked what he ailed. Have I not (quoth he) good cause to weep, that being as there are an infinite number of worlds, I am not yet the lord of one?

According to some ancient Greek philosophers, there were innumerable worlds. Today, we would call this a multiverse filled with infinite parallel universes. Against this idea, other philosophers, notably Aristotle, argued that there was and only could be one world or universe.

---

[1] Plutarch, "Of the Tranquillity and Contentment of Mind."

According to Aristotle, heavy elements fell towards the center of the world, while light elements rose away from that center. If there was another world, it seemed reasonable that it would be made of the same elements. But if there were heavy elements, like earth, in that other world, they ought to be falling towards the same center of the universe as the elements in our universe. Aristotle considered but rejected the possibility that elements might fall towards the closest center. Consequently, there could be no other center and, thus, no other world around that center.

Furthermore, Aristotle argued that all elements must have their natural place within the world. The natural place of heavy elements was at the center of the world. The natural place of light elements was in the atmosphere. The natural place of aether, which moved in a circular fashion, was in the upper heavens. But these were all the possible forms of natural movement, and thus there could not be any further elements. Outside of the heavens could not be the natural place for any elements, and thus it must not exist. Consequently, there could not be anything, even an empty void, outside of the world.

In classical science's conception, the world was a series of concentric spheres. The innermost spheres, distinct from the heavenly spheres, corresponded to the elements: earth, water, air, and fire. Next, there were a series of spheres corresponding to the sun, moon, and planets. Variations of the system included differing additional spheres beyond these. Some considered the outermost sphere to contain the fixed stars. Others considered a starless sphere, or primum mobile, to be the outermost sphere of the heavens. It is thus the highest heavens or the heaven of heavens.

The term "heaven" could be used either to refer to an individual sphere making up the world or the world as a whole. As such, an author referring to multiple heavens might mean either multiple spheres making up the world or multiple worlds. Care must be taken to determine which kinds of heavens the author had in mind.

The Biblical Discussion

Christians believe in a supercelestial realm beyond the stars where God and his angels reside. But this created a conflict with Aristotle's claim that there could not be anything beyond the heavens. Nevertheless, Aristotle's argument could be reconciled with the Christian supercelestial realm if this realm was identified with the outermost sphere of the heavens. This places the supercelestial heaven as the uppermost part of our heavens, beyond even the fixed stars but one that we cannot actually see.

The Genesis 1 creation account refers to the creation of the heavens before the rest of the days of creation:

> In the beginning, God created the heavens and the earth. (Genesis 1:1 ESV)

However, it also describes it as being created on the second day of creation:

> And God made the expanse and separated the waters that were under the expanse from the waters that were above the expanse. And it was so. And God called the expanse Heaven. And there was evening and there was morning, the second day. (Genesis 1:7–8 ESV)

Did God create the heavens twice? Historically, many Christians took this as God's creating two distinct heavens. The outermost sphere, or a region beyond the spheres, could plausibly have been created prior to creation week. Possibly, this is the heavens referred to in Genesis 1:1. The visible heavens that we observe were then created on the second day. The firmament, or expanse, created on the second day corresponds to some or all of the other heavens. In Genesis 1:2, the text describes the earth as being formless and void but says nothing about the heavens. This would make sense if the heavens created here were the realm of God, fully finished and filled with angels.

However, this reading requires understanding the word "heavens" to be referring to two different entities within the same passage. It may be that Genesis 1:1 does not describe a separate act of creation but is rather an introductory summary of the creation account. Alternatively, Genesis 1:1 may refer to the creation of either the space that would become the heavens or the material from which the heavens would be created. In this interpretation, it refers to the whole universe as the heavens and the earth, without intending to claim that the heavens or the earth had yet been created.

Several verses differentiate between the heavens and the "heaven of heavens":

> Behold, to the LORD your God belong heaven and the heaven of heavens, the earth with all that is in it. (Deuteronomy 10:14 ESV)

> But will God indeed dwell on the earth? Behold, heaven and the highest heaven (heaven of heavens) cannot contain you; how much less this house that I have built! (1 Kings 8:27 ESV)

> You are the LORD, you alone. You have made heaven, the heaven of heavens, with all their host, the earth and all that is on it, the seas and all that is in them; and you preserve all of them; and the host of heaven worships you. (Nehemiah 9:6 ESV)

References to the heaven of heavens are readily understood as speaking of a realm that is beyond the heavens. There is very little said about this realm, and it may simply refer to a vague sense of a reality beyond the visible heavens. However, they could also be understood as referring to the highest of the heavens or the outermost sphere. Nevertheless, such a reading is somewhat awkward since the texts seem to imply that the heavens and the heaven of heavens are distinct places.

Jesus' ascension is described as "through and above" the heavens:

> He who descended is the one who also ascended far above all the heavens, that he might fill all things. (Ephesians 4:10 ESV)

> Since then we have a great high priest who has passed through the heavens, Jesus, the Son of God, let us hold fast our confession. (Hebrews 4:14 ESV)

Both of these verses indicate that Jesus ascended past all of the heavens and not merely into the heavens. This is difficult to reconcile with the idea that Jesus ascended into the outermost sphere of the heavens. It could be argued that this refers only to the visible portion of the heavens, up to and including the sphere of the stars, but not the outermost sphere.

The most enigmatic statement about the heavens is from the Apostle Paul:

> I know a man in Christ who fourteen years ago was caught up to the third heaven—whether in the body or out of the body I do not know, God knows. And I know that this man was caught up into paradise—whether in the body or out of the body I do not know, God knows—and he heard things that cannot be told, which man may not utter. (2 Corinthians 12:2–4 ESV)

Most commentators take Paul to be talking about himself in the third person. He states that he was caught up to the third heaven and into paradise. Generally, this is understood as

143

Paul's visiting or having a vision of the supercelestial realm. Nevertheless, it is unclear what exactly the phrase "third heaven" refers to.

A commonly given interpretation is that the first heaven is the atmosphere, the second heaven is outer space, and the third heaven is God's realm. However, did Paul and his contemporaries think of the atmosphere and space in terms of being the first and second heavens? The Biblical text does not clearly distinguish between the atmosphere and space, let alone number them.

Is Paul referring to one of the multiple heavens of classical cosmology? Perhaps he is claiming to have visited Venus. Probably not. Paul is writing long before Neoplatonic thought would bring this theory to broad acceptance. Furthermore, there is no standard numbering scheme that would indicate which of the heavens Paul is referring to.

An alternative interpretation is that Paul is not referring to a visit to the supercelestial realm at all. Rather, he is referring to a vision or visit to the new heavens and the new earth. In this interpretation, the phrase "first heavens" refers to the universe at the time of the original creation. As Peter says, this first heavens and earth perished in the flood:

> For they deliberately overlook this fact, that the heavens existed long ago, and the earth was formed out of water and through water by the word of God, and that by means of these the world that then existed was deluged with water and perished. But by the same word the heavens and earth that now exist are stored up for fire, being kept until the day of judgment and destruction of the ungodly. (2 Peter 3:5–7 ESV)

We then live in the second heavens and earth, but this heavens and earth will only remain until the day of judgment, when they will be replaced by the third heavens and earth. Another reason to favor this interpretation is that Paul describes the place as παράδεισος (paradeisos) which derives from a Persian word referring to a garden owned by the nobility. It is most likely a reference to the restored Garden of Eden owned by God.

It is certainly possible to read these texts in a way that is compatible with the supercelestial heavens' being the outermost sphere of a series of concentric spheres. However, in most of the passages, it is the less plausible reading. References to the heaven of heavens, passing far above all the heavens, and the third heavens must all be understood in an awkward fashion to be compatible with the supercelestial realm's being the uppermost sphere. Instead, the Biblical text seems to refer to the existence of a supercelestial realm beyond the heavens.

Historical Christians

Origen of Alexandria

Origen distinguishes between the heaven of Genesis 1:1 and the firmament of the second day.[2] Heaven is the supercelestial realm, whereas the firmament is the heavens that we see. Origen refers to Isaiah describing heaven, which he takes to be the supercelestial realm, as God's throne:

> Thus says the LORD: "Heaven is my throne, and the earth is my footstool; what is the house that you would build for me, and what is the place of my rest? (Isaiah 66:1 ESV)

Origen describes this as a spiritual substance that God uses as a throne. He distinguishes this from the corporeal heaven. The spiritual heaven was made in the beginning before the rest of

---

[2] Origen, *Homilies on Genesis and Exodus*, Homily 1.

creation. The firmament, created on day two, is the physical heavens that we can see.

Celsus, a pagan philosopher, argued against Christians who believed in a supercelestial God, that is, a God who resides beyond the heavens.[3] Furthermore, he alleges that this belief derives from a misunderstanding of the works of Plato. Celsus quotes Plato saying, "All things are around the King of all, and all things exist for his sake, and he is the cause of all good things." Compare this to Paul's statement, which invokes a similar idea:

> And he is before all things, and in him all things hold together. (Colossians 1:17 ESV)

Celsus maintains that Christians have misunderstood Plato and thus believe, incorrectly, in a supercelestial God. In contrast, Celsus believes that the Jews, being more sensible than Christians, only believe in a God who resides in the heavens and not beyond them.

Origen responds by arguing that the Old Testament prophets predate Plato and that they already taught a supercelestial God. Consequently, this idea could not have been borrowed from a misunderstanding of Plato. To support this, he quotes the Psalms:

> Praise him, you highest heavens (heaven of heavens), and you waters above the heavens! (Psalm 148:4 ESV)

Origen takes this verse to be referring to a supercelestial realm. This could be because it refers to the heaven of heavens, the presence of water above the heavens, or both.

Origen suggests that Plato might have learned his idea from some Jews or by reading the Old Testament prophets.

---

[3] Origen, "Contra Celsum," Book VI, Chapter 18-19.

He quotes from a book by Plato called *Phaedrus*, referring to a supercelestial realm:[4]

> No poet here below has ever sung of the super-celestial place, or ever will sing in a becoming manner.

Origen suggests that Paul was familiar with these words, and cites Paul's discussion:

> For this light momentary affliction is preparing for us an eternal weight of glory beyond all comparison, as we look not to the things that are seen but to the things that are unseen. For the things that are seen are transient, but the things that are unseen are eternal. (2 Corinthians 4:17–18 ESV)

Origen takes Paul's references to the unseen to be references to a supercelestial realm. Thus Origen concludes that the doctrine of the supercelestial realm is not the invention of later Christians but the teaching of Paul.

Basil of Caesarea

Basil draws attention to the creation of the heavens and the firmament.[5] He argues that this describes the creation of two distinct heavens. However, he suggests that the philosophers of his time would reject this. They would insist that there could only be one heaven. There could only be one heavens circling around one center of the universe.

Basil points out that other philosophers thought there were multiple or even infinite heavens. This is a continuation of the argument Basil has previously made, pitting the philosophers of his day against each other. Their disagreements show that their words have been nothing but empty wind.

---

[4] Plato, "Phaedrus," 247C; Origen, "Contra Celsum," Book VI, Chapter 19.
[5] Basil of Caesarea, "Hexaemeron," Homily III.

Basil points out that the same cause can bring forth multiple bubbles. As such, it seems that an omnipotent God is quite capable of forming multiple heavens. He finds their argument of impossibility simply ridiculous.

Basil refers to Paul's visit to the third heavens as indicating that there is not only a second heaven, but a third heaven. He also points to the psalmist's reference to the heaven of heavens, which he takes as an indication that there are multiple heavens.

Basil points to the fact that classical cosmology divided the heavens into seven circles. These are the ones corresponding to the sun, moon, and five classical planets. Given this, how can it be strange that there could be multiple heavens? The philosophers, despite their argument that multiple heavens are impossible, still believed in some idea of multiple heavens.

Elsewhere, Basil asks why the world was dark when it was created—why was there darkness over the face of the deep (*Hexaemeron*, Homily II, Section 5)? Darkness is inferior to light and thus ought to have been created after it. Furthermore, the angels, heavenly hosts, spirits, etc. that already existed must have dwelt in light and not darkness.

Basil insists that no one could disagree that celestial light is one of the rewards promised to the righteous. To support this, he quotes two verses:

> The light of the righteous rejoices, but the lamp of the wicked will be put out. (Proverbs 13:9 ESV)

> giving thanks to the Father, who has qualified you to share in the inheritance of the saints in light. (Colossians 1:12 ESV)

In contrast, those who are condemned are sent into outer darkness,

> while the sons of the kingdom will be thrown into the outer
> darkness. In that place there will be weeping and gnashing
> of teeth. (Matthew 8:12 ESV)

As such, Basil concludes that the supercelestial heaven must be filled with light. But if so, why was the world dark?

Basil argues that the heaven created in Genesis 1:1 was some sort of body enveloping what we know as our world. This body blocks the light of the realm of God. Consequently, everything inside is dark.

Basil compares this to being inside a tent at midday. By entering the tent, it can suddenly be dark no matter how bright it is outside. He suggests that this is the nature of the primordial darkness. It was dark because the light was blocked by the newly created heavens.

The interesting point for our purposes is that Basil conceives of the supercelestial realm as part of our physical world. In fact, the light from that realm would be visible from earth if it were not blocked by the heavens.

Basil argues that there are multiple heavens in explicit opposition to philosophers who argued that there could only be one heaven. He insists that the Scripture teaches the existence of these multiple heavens. He argues that the philosophical opposition to multiple heavens is unfounded.

Ambrose of Milan

Ambrose asks whether the firmament created on day two is distinct from the heavens created in Genesis 1:1.[6] Furthermore, how many heavens are there? He notes that some maintain that there can be only one heaven. This is because there is one ὕλη (hyle) or matter. He is referring to Aristotle's argument, which claimed that all matter was contained in one heaven and

---

[6] Ambrose of Milan, "Hexameron," Book II, Chapter 2.

there could be no other. He notes that some philosophers also argued the opposite, that there were countless heavens.

Ambrose rejects the argument of impossibility. If man can make multiple objects from the same material, surely God can do so as well. He supports his point with a verse speaking of God making the heavens:

> For all the gods of the peoples are worthless idols, but the LORD made the heavens. (Psalm 96:5 ESV)

He further points to two passages that speak of God's being able to do whatever he wants:

> Our God is in the heavens; he does all that he pleases. (Psalm 115:3 ESV)

> And he said, "Abba, Father, all things are possible for you. Remove this cup from me. Yet not what I will, but what you will. (Mark 14:36 ESV)

Ambrose insists that we must accept the existence of the second and third heavens. As support, he points to Paul's visit to the third heaven and the psalmist's references to the heavens of heavens. Ambrose further suggests that the philosophers took the idea of multiple heavens associated with the sun, moon, and planets from the psalmist. Ambrose follows Basil in making many of the same arguments and arguing for the existence of multiple heavens.

Augustine of Hippo

Augustine interprets references to the heaven of heavens in different ways in different places. In the *Confessions*, he suggests that the heaven of heavens is a realm beyond the loftiest heavens that we can see.[7] However, in *On the Literal Meaning*

---

[7] Augustine, "Confessions," Book XII.

of Genesis, he rejects this interpretation.[8] Instead, he interprets the heaven of heavens as referring to the realm of stars.[9]

The final book in *On the Literal Meaning of Genesis* is devoted to understanding Paul's reference to the third heaven. He summarizes his interpretation:[10]

> It seems that we are right, then, in understanding the first heaven in general as this whole corporeal heaven (to use a general term), namely, all that is above the waters and the earth, and the second heaven as the object of spiritual vision seen in bodily likenesses (as for instance, the vision seen by Peter in ecstasy when he saw the dish let down from above full of living creatures), and the third heaven as the objects seen by the mind after it has been so separated and removed and completely carried out of the senses and purified that it is able through the love of the Holy Spirit in a mysterious way to see and hear the Divine Word through whom all things are made.

Augustine's interpretation is unusual.

John Chrysostom

Chrysostom rejects the existence of many heavens.[11] He insists that it is an idea derived from human reasoning rather than Scripture. Most likely, he rejected not the existence of a supercelestial realm but the division of the heavens into a series of concentric spheres. As we learned previously, Chrysostom rejected the theory that the heavens were a sphere and thus would also reject the idea that the heavens were a series of concentric spheres. We will see that this makes the most sense out of what Chrysostom says.

---

[8] Augustine, *The Literal Meaning of Genesis. 1,* Book 1, Chapter 9, Section 17.

[9] Augustine, Book II, Chapter 18.

[10] Augustine, Book 12, Chapter 34, Paragraph 67.

[11] Chrysostom, *Homilies on Genesis,* Homily IV.

In a prior homily, Chrysostom described the Genesis 1:1 heavens as bright and finished,[12] in contrast to the invisible and unfinished earth. But this is prior to the creation of light or stars, so the corporeal heavens would be neither bright nor finished. However, if he is referring to a supercelestial realm, it would make sense that he thought they were created in a bright, finished state.

He argues that Moses does not speak of the creation of multiple heavens. But as many other authors had pointed out, Moses does appear to speak of the creation of multiple heavens. However, this makes sense if Chrysostom is focused on the division of the heavens into multiple spheres. He takes the second day to be the creation of the visible heavens, and since only one heaven was created, the visible heavens cannot be made up of multiple heavens. If the heavens were made up of multiple spheres, Moses would have described the creation of multiple heavens.

Chrysostom considers the objection that the psalmist refers to the heaven of heavens. However, his response focuses entirely on the point that the word "heavens" is plural, not on how the heaven of heavens differs from the heavens. Chrysostom points out, correctly, that the Hebrew language always uses the word heavens in the plural. This is akin to English words like clothes, information, or furniture which only exist in plural forms. As such, the fact that the Scriptures refer to the heavens rather than heaven does not demonstrate that there are multiple heavens. Chrysostom insists again that if there were multiple heavens, Moses would have spoken of their creation.

---

[12] ,Chrysostom, Homily 2 on Genesis, Paragraph 12.

Cosmas Indicopleustes

Cosmas indicates that the creation of the firmament on the second day is the creation of a second heaven.[13] He thinks it is made from water and has a similar appearance to the first heaven, created in Genesis 1:1. Cosmas argues that everything from the earth up to the firmament is the first of two places in this world. From the firmament upwards is the second place, the Kingdom of Heaven, or supercelestial realm.[14] Throughout his book, he repeatedly insists that God only made these two places and no others.

Cosmas rejects the existence of the spheres corresponding to the sun, moon, and planets.[15] Cosmas asks Christians which of these multiple heavens they think Jesus ascended into. Did he ascend into the sphere of the moon, Mars, Saturn, or one of the other planets? Which heaven do these Christians one day hope to reach? What is the point of all of these different heavens? Pagan beliefs about the heavens, Cosmas insists, are not compatible with Christian belief.

Cosmas's argument seems strange. Nobody thought they were going to ascend to one of the visible heavens. But Cosmas appears to think those who accepted classical cosmology took the system of concentric spheres to be the entirety of reality. Thus, there was no supercelestial realm. Consequently, the spheres were the only places one could conceivably ascend into.

Cosmas has difficulty accounting for Paul's visit to the third heaven.[16] In his scheme, the first heaven is the supercelestial realm. The second heaven is the firmament. There is no room for a third heaven in this scheme. Cosmas argues that Paul

---

[13] Indicopleustes, *Christian Topography*, Book III.

[14] Indicopleustes, Book IV.

[15] Indicopleustes, Book IV, Note 2.

[16] Indicopleustes, Book VII.

meant that he was carried up two thirds of the way to heaven. This is not a grammatically plausible reading of Paul.

Bede

Bede comments on:

> The earth was without form and void, and darkness was over the face of the deep. And the Spirit of God was hovering over the face of the waters. (Genesis 1:2 ESV)

Bede takes the fact that only the earth is described as being without form and void as indicating that this does not apply to the heavens, which were created in Genesis 1:1.[17] These highest heavens are not part of the changeable world. This is distinguished from our heaven, the firmament, where the heavenly bodies reside. The highest heaven, which we cannot see, was not created formless and void. Instead, it was created and filled with the blessed hosts of angels.

John of Damascus

John takes heaven, in the broadest sense, to include all of creation, including the abode of angels.[18] God, however, cannot be contained in heaven, as he is the omnipresent creator.

Given the various references to heaven, the heaven of heaven, the heavens of heavens, and the third heaven, John indicates acceptance of the starless sphere. This starless sphere, or primum mobile, was the outermost sphere of the heavens. It is thus the highest heavens or the heaven of heavens. Furthermore, John suggests that the foreign philosophers who proposed this sphere actually borrowed the idea from Moses.

John identifies the heaven of heaven as the first heaven, which is above the firmament. The firmament is also called a

---

[17] Bede, "Commentary on Genesis."

[18] John of Damascus, "An Exposition of the Orthodox Faith," Book II, Chapter 6.

heaven, and thus the second heaven. Furthermore, the Scriptures often speak of the air as being the heavens. In particular, birds are said to fly in the heavens even though they fly in the air. As such, there are three heavens: the air, the firmament, and the heaven of the heavens, just as the Apostle Paul indicated.

Remigius of Auxerre

Remigius takes Genesis 1:1 as referring to the creation of the supercelestial spiritual heavens.[19] The physical heavens, or firmament, were created later.

Thomas Aquinas

Aquinas discusses whether there is only one heaven.[20] Aquinas invokes the psalmist's call upon the heavens of heavens to indicate that there are multiple heavens. In the third objection, he accepts Aristole's argument that there can only be one heavens. He takes these heavens to be the multiple concentric spheres of the classical cosmology. He goes into detail, discussing the division of the heavens into spheres corresponding to the elements and planets.

Aquinas interprets Basil and Chrysostom as being in agreement with his view. Basil had insisted on the existence of multiple heavens. Aquinas thought that these different heavens were the different concentric spheres of classical cosmology. However, it is clear in Basil's discussion that he differentiates those spheres from the other heavens that he insists exist. Basil was insisting on the existence of realms beyond the heavens.

Chrysostom had rejected the existence of multiple heavens. Aquinas thinks that he was rejecting the existence of multiple distinct heavens or universes. However, as argued earlier,

---

[19] Remigius of Auxerre, "Exposition on Genesis."

[20] Aquinas, *The Summa Theologiæ of St. Thomas Aquinas*, The First Part, Question 68, Article 4.

Chrysostom was rejecting the multiple spheres of classical cosmology. Most likely, he accepted the existence of a supercelestial realm beyond the heavens.

Aquinas argues that the uppermost sphere of the heavens, the empyrean heaven, was created at the same time as matter.[21] The angels were created in this sphere.[22] He points out that the second creation of the heavens can be understood in a number of ways.[23] He attributes to Chrysostom an interpretation in which the first verse is a summary of the entire creation, with the rest of the passage giving a more detailed account. The first creation might have simply created the raw material from which the heavens were constructed on the second day. The second heavens might be merely the air around the earth, or the heavens excluding the empyrean heaven. Aquinas rejected prior authors' arguments for the empyrean heaven as not cogent, partially because of the different ways to understand the two different creations of the heavens. Instead, he argues that it is fitting that there was a part of the world that was free from corruption as a beginning of the future glory of the new heavens and the new earth.

Martin Luther

Luther suggests that in Genesis 1:1, God created the rude unformed heavens and earth.[24] However, on the second day, God formed the heavens out of the original nebulous creation. It was on that day that God created the heavens as we know them, except for the heavenly bodies, which would not be created until the fourth day.

---

[21] Aquinas, The First Part, Question 66, Article 3.

[22] Aquinas, The First Part, Question 61, Article 4.

[23] Aquinas, The First Part, Question 68, Article 1, Objection 3.

[24] Luther, *Luther On The Creation: A Critical and Devotional Commentry on Genesis*, 1:Part II, Section I.

Luther states that the word heaven is often used to refer to the horizon, or that portion of the sky that can be seen from any place. The heaven in Germany is thus different from the heaven in other places because a different portion of the sky is visible in each place. Furthermore, he suggests that the Bible uses the phrase "heaven of heavens" to refer to the whole heaven and not just the part visible from a particular place.

John Calvin

Calvin thinks that the firmament is the area below the clouds. We will look at that in more detail when we consider whether there are waters above the heavens.

Calvin comments upon Paul's visit to the third heaven.[25] Calvin takes Paul to be referring to a supercelestial heaven. However, the third heaven refers to the highest and most complete portion of that heaven. He suggests that those who are "superior to others in knowledge" attain a higher degree in heaven, suggesting that few reach the innermost heaven.

Calvin takes the reference to the heaven of heavens as referring to the system of spheres surrounding the earth.[26] As Calvin points out, eclipses and other observations demonstrate that the stars are farther away than the planets, and the planets each have their own orbits. According to Calvin, the psalmist is praising the excellency of this contrivance. He does not suggest that there are multiple heavens, but simply extols the wisdom of God in creating the heavens and giving each heavenly body its own position and station.

---

[25] Calvin, *Calvin's Commentary on the Bible*, Commentary on 2 Corinthians 12:1-5.

[26] Calvin, Commentary On Psalm 148.

Conclusions

Historical Christians

A consistent tendency is for historical Christians to incorporate the supercelestial realm into their cosmology. Basil thinks of it as a realm filled with light surrounding our world. Likewise, Cosmas insists that the supercelestial realm is the region above the firmament. Aquinas takes it to be the outermost sphere in the system of spheres surrounding the earth. This heaven is not so much beyond the heavens as a part of them.

Origen, Basil, and Ambrose defend the existence of the realm against philosophical objections. They insisted that God was able to create additional heavens beyond what we saw. Chrysostom and Cosmas reject the existence of a series of concentric spheres. Since these two individuals rejected the claim that the heavens were a sphere, this is not surprising.

The two creations of the heavens in the Genesis 1 creation account were frequently cited to support the idea of a supercelestial realm. The same is true of references to the heaven of heavens. However, the two most recent authors, Luther and Calvin, take alternative interpretations of this phrase. There is essentially no agreement on how to understand Paul's reference to the third heavens.

Modern Science

Aristotle's argument for there being only one system of concentric spheres would be overturned by the arrival of heliocentrism. Under heliocentrism, there were no longer a series of concentric spheres making up different layers of the heavens. Aristotle's argument rested on his conception of the nature of gravity: things move towards, away from, or around the center of the universe. Since his conception is now clearly false, his argument is moot.

In some sense, every moon and planet is its own heavens in the sense conceived of by Aristotle. Each one has a center of gravity toward which nearby objects fall. Thus, not only are there other heavens, but we can see them. Modern man has even visited one of them. Furthermore, modern science is very open to the possibility of a multiverse of parallel universes.

But what is beyond the farthest stars? We cannot be certain because light takes time to travel. Beyond a certain distance, we simply cannot directly detect in any way what might be present. Furthermore, due to the expansion of the universe, anything beyond that distance is moving farther and farther out of sight. Many scientists think there is simply an infinite universe of additional stars.

Evaluating the Response

In this case, historical authors have not generally fared well. The concentric spheres of classical science turned out to be incorrect, but not in a way that agreed with the authors who rejected them. However, those who identified phrases like the heaven of heavens as referring to these concentric spheres also fared poorly. What went wrong?

Chrysostom and Cosmas both rejected the multiple spheres because the creation account does not mention them. This is not a very compelling argument, because there are many details of creation that are not mentioned in that account. However, this is consistent with these authors' prior rejection of classical gravity and the spherical heavens. They do not have good exegetical reasons for this position, and it appears they began with their rejection of classical cosmology and try to find that in the text.

Other authors accepted the multiple spheres and managed to fit them into the text with references to the heaven of heavens. However, a more careful investigation would make it clear that these authors would not have known about classical cosmology. As such, it is clear that they cannot have been referring to it. These authors were reading the scientific ideas of their day back into the Biblical text.

Despite coming to opposite conclusions, both are based on the same mistake. Instead of going back to the source to understand what the text is seeking to convey, they read their scientific ideas into the text. They started with different scientific ideas and thus read different ideas into the text, but nonetheless, they were all reading their own ideas into the text. Failing to return to the source led them into error.

Another theme is the identification of the supercelestial heavenly realm as another sphere in the system of spheres, or at least a physical realm adjacent to ours. This identification has fared poorly, as it is now clear this sphere does not exist. However, as discussed in the introduction, various Biblical passages suggest that the supercelestial heaven is not simply the farthest part of our cosmology. Certainly, there was no Biblical basis for the incorporation of the supercelestial realm into classical cosmology. These authors were overly influenced by the science of their day.

At the same time, this illuminates how historical authors regarded the Biblical text. They were concerned to understand the double creation of the heavens, references to the heaven of heavens, and Paul's reference to the third heaven. They saw the Scriptures as having authority, even on questions of cosmology. Nobody rejected the Scripture's ability to speak to questions of cosmological science.

Despite the errors in handling this issue that we have iden-tified, there is something to be encouraged about. The Church considered and rejected the arguments of those who rejected classical cosmology. They were able to judge that it was not based on good exegesis and move forward. Likewise, the Church has moved on from interpreting the Scriptures in terms of classical cosmology. Despite historical authors interpreting it that way, it is no longer an issue today.

This happens because the Church went back to the source. When they evaluated the arguments of those opposed to classical cosmology, they found them to not be rooted in Scripture. Likewise, when classical cosmology was replaced, the Church reevaluated their understanding of Scripture, seeing that it simply did not teach classical cosmology. Returning to the source of what the Scriptures actually said was the solution to these errors.

Chapter 9

Are There Waters Above the Heavens?

The Biblical Discussion

The second day of the Genesis 1 creation account states:

> And God said, "Let there be a [*raqia*] in the midst of the waters, and let it separate the waters from the waters." And God made the [*raqia*] and separated the waters that were under the [*raqia*] from the waters that were above the [*raqia*]. And it was so. And God called the [*raqia*] Heaven. And there was evening and there was morning, the second day. (Genesis 1:6–8 ESV)

Due to uncertainty about its meaning, the Hebrew word, *raqia*, is variously translated as "expanse," "space," "firmament," "vault," "dome," or "horizon." Most of the uses of this word are in the Genesis 1 creation account. Ezekiel refers to it in a couple of passages (Ezekiel 1:22–26, Ezekiel 1:10). However, those passages do not give much information and are about a likeness of the *raqia*, not the *raqia* itself. A few other passages use the word (Psalm 19:1, Psalm 150:1, Daniel 12:3); however, these references do not elucidate the meaning of the word.

In addition to the *raqia*, the passage speaks of waters above the *raqia*. One other passage also refers to these waters:

Praise him, you highest heavens, and you waters above the heavens! (Psalm 148:4 ESV)

But what are these mysterious waters that are beyond the heavens? The *raqia* and the waters above are two of the most enigmatic and debated aspects of the creation account.

There are three key points that pose a challenge to most proposed interpretations. Firstly, the *raqia* is explicitly named the heavens. Any plausible interpretation must understand the *raqia* as essentially the same thing as the heavens. Secondly, the creation account describes the creation of the natural world as it would have been understood by the ancient Israelites. It does not address the creation of things they would not have known about, such as bacteria. It also does not discuss the creation of things beyond the natural world, such as angels. Any interpretation that takes the text to refer to the creation of things outside of the natural world or otherwise unknown to us is highly suspect. Thirdly, the purpose of the *raqia* is stated to be to divide the waters. Any interpretation has to make sense of this purpose.

A common modern interpretation is that the waters above refer to clouds. This is immediately questionable because this would be a highly atypical way to refer to clouds. Furthermore, it requires identifying the *raqia* with only the portion of the heavens between the earth and the clouds. This is particularly problematic because the sun and moon are explicitly described as being created in the *raqia*. The sun and moon are clearly not between the earth and the clouds. Furthermore, how does it make any sense to describe the air as serving to separate the oceans from the clouds?

An alternative is to argue that there is a body of water beyond the farthest stars. The *raqia* thus includes the entirety of what we now call outer space. However, this has the creation account referring to the creation of a body of water that we would otherwise know nothing about. Furthermore, how does it make any sense to describe the purpose of space as being to separate the waters on earth from distant waters beyond the stars?

Another alternative takes the text to be teaching a primitive flat-earth cosmology. In this interpretation, the *raqia* is a dome sitting on a flat earth holding back a cosmic ocean. It is argued that this was a common cosmological conception in the ancient world and thus would have been describing aspects of the world recognizable to the ancient person. However, on this interpretation, the *raqia* is not the heavens; rather, the *raqia* is one part of a more complex cosmology. Furthermore, how does it make any sense to think of this dome as having the purpose of separating waters from waters?

None of the commonly given interpretations of this passage are very compelling. It is particularly telling that in none of these interpretations does it actually make much sense to describe the *raqia* as separating the waters from the waters. Remember that we've previously seen water in the creation account:

> The earth was without form and void, and darkness was over the face of the deep. And the Spirit of God was hovering over the face of the waters. (Genesis 1:2 ESV)

Contextually, we would expect the text to be referring to these waters that have already been introduced. As such, the text is speaking of the division of that original body of water. It is not speaking of bodies of water that still exist, but rather of the body of water that existed during creation and was divided at that time.

The words translated as "above" and "below" are not the standard Hebrew words for above or below. Instead, they both have the added prefix מ (*mem*), which means "from." The one other reference from the Psalms to the waters above also uses this prefix. As such, the text speaks of the waters *from* below the *raqia* and the waters *from* above the *raqia*. These words are usually used to describe where something was before it was removed. The implication is that these waters are no longer where they were.

It follows naturally from this interpretation that the waters below would become the land and seas on the next day of creation, while the waters above would become the heavens. This would mean that the *raqia* is made from the waters above. However, most have understood the *raqia* to be something placed in between the waters above and the waters below. But does the text actually indicate that?

The text states, "let there be a *raqia* in the midst of the water." The Hebrew word translated as "midst," *tavek*, never means between. Rather, it means among, within, or in the middle. As such, it merely indicates that the *raqia* is created in the water, probably from water.

The *raqia* is described as dividing the waters above and the waters below. However, the Hebrew word here, בָּדַל (*badal*), refers to the act of dividing, cutting, or partitioning something. Rarely, it can refer to a physical object such as a veil (Exodus 26:33) or wall (Ezekiel 42:20) that marks such a division; however, it usually does not. In the creation account, the same word is used of the sun and moon, dividing the day and the night. This word does not actually suggest that the *raqia* is some sort of barrier between the waters above and below.

The text speaks of the waters above the *raqia*. However, the Hebrew preposition here, *al*, has a much wider range than the English word "above." It is more similar to the English preposition "on." Things can be on other things in a variety of senses: a shoe can be on a foot, a poster can be on a wall, and a book can be on a table. In this case, the text is referring to the waters, which were in the upper position now taken by the *raqia*.

Later passages refer back to the creation of the heavens as a circle being inscribed on the face of the deep, suggesting that they read it this way:

> He has inscribed a circle on the face of the waters at the boundary between light and darkness. (Job 26:10 ESV)

> When he established the heavens, I was there; when he drew a circle on the face of the deep, (Proverbs 8:27 ESV)

Putting all of this together, I think there is a good case for this interpretation. On the second day, God divided the waters into the lower waters, which became the earth and seas, and the upper waters, which became the *raqia*, or heavens. The *raqia* is simply the heavens. It does not refer to any exotic or unknown entities. This is the only interpretation that makes sense of *raqia* being described in terms of a division of waters from waters. The actual Hebrew words support this interpretation and other Biblical passages appear to interpret it this way.

The Septuagint translated this text in line with the conception of a barrier holding up waters. It translated *raqia* as στερέωμα (stereōma) which derives from στερεός, which means "firm" or "solid." The word "above" in the "waters above" is translated as ἐπάνω (epanō), which loses the meaning of removal and more explicitly means above. The Vulgate translated *raqia* as *firmamentum* and above as "super" which both have similar meanings to the Greek words. This was followed by many influential English translations of the Bible, leading to the translation familiar to many people: firmament.

The Scientific Question

Classical science considered the universe to consist of a series of concentric spheres surrounding the earth. Scholars between the scientific revolution and the 1980s thought that classical cosmology considered these spheres to be hard and impenetrable. The idea of hard and impenetrable spheres surrounding the earth has obvious similarities to that of some sort of firmament. However, modern scholars have concluded that classical cosmology did not, in fact, believe that the spheres were hard and impenetrable .

Classical cosmologists did describe these spheres as solid. While the word solid today often refers to something hard and impenetrable, this is not the only meaning. If a statue is "solid gold," this means that it is gold all the way through, with the inside being neither empty nor made of other material. In the field of solid geometry, solid refers to three-dimensionality. Because the word "solid" can mean different things, we must be careful to ascertain how it is being used in any particular context.

Classical cosmology rejected the idea that there could be a body of water beyond the heavens. A heavy element like water ought to fall down onto the earth. Within classical science, this was thought to apply no matter how far away from the earth the water was to be found. Even waters in the distant reaches of the heavens ought to fall down to earth. The proper place for water was on the earth, not in some region beyond the air. It did not make any sense that there could be a second, proper place for water.

The obvious explanation would be that water is prevented from falling down by the firmament. But this raised its own set of difficulties. Anything solid enough to hold up the waters would have to have been heavy and thus should fall to earth itself. Furthermore, given that the heavens were a spinning sphere, water could not very well remain on top of this sphere, even if it were hard.

However, the series of concentric spheres suggested a way to reconcile the apparent Biblical teaching with classical cosmology. The firmament and the waters could be easily understood as referring to two of these spheres. Either they referred to some of the spheres already known to classical science or there were additional spheres unknown to classical science that the Biblical text was referring to. As such, while the idea of a firmament and water above conflicted with classical science, there was an obvious direction to take in attempting to reconcile them.

Historical Christians

Most of the historical authors we will consider used either the Septuagint or the Vulgate, which translated the word *raqia* as either *stereōma* or *firmamentum*. English translations of these authors typically render the word as firmament, and we will follow that practice as we discuss their handling of the text.

Theophilus of Antioch

Theophilus indicates that half of the waters were taken up to the firmament, or heaven that we can see.[1] These waters provided rain, showers, and dew. The water left on earth provided for rivers, fountains, and seas. He says that the waters were taken *to* the firmament rather than *above* the firmament, suggesting that he sees these waters as residing in the heavens, rather than in a supercelestial position.

---

[1] Theophilus of Antioch, "To Autolycus," Book II, Chapter 13.

Origen of Alexandria

Origen identifies the firmament as the physical heavens, as opposed to the heavenly realm of God.[2] He identifies the waters below as the abyss, the dwelling place of the devil and his angels. He draws from Genesis 1:1 which, in the Septuagint, refers to darkness being upon the abyss, and Luke 8:31, in which the demons beg not to be sent to the abyss. His identification of the waters above is less clear, although he refers to them as spiritual and celestial waters. He calls upon his hearer to participate in these spiritual waters by applying their minds to lofty and exalted things.

But why are the physical heavens called a firmament? Origen says that every physical object "is, without doubt, firm and solid." He does not elaborate, but may be referring to the idea that even though the firmament is not as firm and solid as the earth, it still has a definite substance to it and is much firmer and more solid than a spiritual substance. He also states that heaven is called the firmament because it divides between the waters above and the waters below. Again, he does not elaborate but may be referring to the idea that the firmament has the strength to keep the waters above from falling to the earth.

Basil of Caesarea

Basil takes firmament to refer to the physical heavens, distinguishing them from the previously created supercelestial realm.[3] He compares other passages from the Septuagint that use the same word, *stereōma*:

---

[2] Origen, *Homilies on Genesis and Exodus*, Homily on Genesis 1.
[3] Basil of Caesarea, "Hexaemeron," Homily 3.

The LORD is my rock (*stereōma*) and my fortress and my deliverer, my God, my rock (*stereōma*), in whom I take refuge, my shield, and the horn of my salvation, my stronghold. (Psalm 18:2 ESV)

When the earth totters, and all its inhabitants, it is I who keep steady (make *stereōma*) its pillars. *Selah* (Psalm 75:3 ESV)

Praise the LORD! Praise God in his sanctuary; praise him in his mighty heavens (*stereōma*)! (Psalm 150:1 ESV)

Only the last of the passages is a translation of the Hebrew word *raqia*. While others use the same Greek word, they are translations of different Hebrew words. He argues that the Scripture uses the word *stereōma* to refer to "all that is strong and unyielding." He goes on to say that the firmament is not "in reality a firm and solid substance," but rather is only called a firmament by comparison with the substance of other bodies in the heavenly realms.

Basil says that the earth was originally surrounded by a large mass of water suspended in the air. He argues that this mass of water was necessary in order to counteract the heat of the sun. Over time, the sun destroys these waters, so God filled the world with enough water to last until the end of the world. For now, the waters above provide a source for rain and dew.

Basil considers an objection: If the firmament is a sphere, how could it be holding up water? A convex shape, such as a sphere, is highest in the middle, and thus the waters would be expected to run off. Basil responds that while the inside of the heavens, which we can see, is curved, the exterior might not be. He points to the example of buildings that are internally dome shaped and yet have a flat roof.

Basil explicitly and vociferously disagrees with those who suggest that the waters above are merely symbols for spiritual powers. This is often understood to be referring to Origen. However, this view does not exactly line up with Origen's statements. Origen did not state that the waters were spiritual powers, but that they were spiritual substances. The spiritual powers lived in these waters but were not the waters themselves. Nevertheless, either Basil is responding to Origen or someone heavily influenced by Origen's ideas. Basil has sharp words for these ideas, calling them old women's tales. He insists that by water, water is meant.

Ambrose of Milan

Ambrose identifies the firmament as the physical heavens.[4] Ambrose discusses the doubts that might arise about the waters above. It would seem that a solid body like the firmament could not be at rest between the liquid bodies of the upper and lower waters. Furthermore, water, by its nature, collects in one place and thus would not naturally separate into upper and lower waters. According to Ambrose, because God created nature, he is entitled to override these principles and make the waters do what he wishes.

Ambrose insists that there are waters above the firmament despite all objections. He says about the Scriptural teaching about the waters, "What is clearer than this?" Like Basil, he addresses the objection that water would run off of the sphere by arguing that the sphere might be a different shape on the exterior. He also argued that God had created the large amount of water to balance the extreme heat of the sun.

Ambrose also makes an argument for the possibility of water in the heavens by asking: Where does the rain come from? It does not matter how one accounts for rain; one way or the other, water has to get up and stay in the heavens. This

---

[4] Ambrose of Milan, "Hexameron," Book Two.

proves it is possible for water to be up in the heavens. As such, the objection that it would be impossible for water to be in the heavens fails.

Ambrose also argues that if the earth can be kept immobile in the center of the universe, surely it must also be possible for waters to be kept in the heavens. Alternately, Ambrose suggests that the rotation of the heavens might keep the waters aloft.

Augustine of Hippo

Augustine addresses an argument that Christians could not accept the heavens being in motion, because the heavens were a firmament.[5] He writes that he does not have further time to go into these questions. However, he insists that the heavens could be in motion and still be a firmament in the sense that they are solid and provide an impassable boundary. He also says that if the heavens turned out to be immovable, it would still be possible to explain the apparent motion of the heavenly bodies.

In an earlier commentary, Augustine argued that the waters in the creation account were formless matter.[6] Thus, he takes the division of waters as dividing the corporeal matter of our world from the incorporeal matter of the invisible world. He suggests that the incorporeal world is said to be above, not because it is physically above but because it is more beautiful than our world. Nevertheless, he is quick to say that we should not rashly affirm things about it because it is obscure and remote from our senses. He insists that even before we understand it, we must believe it.

---

[5] Augustine, *The Literal Meaning of Genesis. 1*, Book 2, Chapter 10.

[6] Augustine, "Two Books on Genesis against the Manichees," Book 1, Chapter 11, Section 17.

In that book, Augustine took the view that the waters above are not part of of the physical world. However, in *The Literal Meaning of Genesis*, he assumes that they are part of the physical world. Augustine considers the objection that there cannot be waters above the heavens because the waters would fall down to the earth.[7] A possible response would be that God could keep the waters above the heavens if he wanted to. Augustine emphatically rejects this answer. Yes, God could keep the waters above heaven if he wanted, just as he could keep oil below water. However, we seek to know how God made the world and whether he made the world in a way that allows water to reside above the heavens.

Augustine describes an explanation given by a "certain commentator." The commentator points out that the air is often referred to as the heavens, especially in the Scriptures. In particular, he points to the many references to the birds of the heavens (often translated as "birds of the air"), as well as Jesus speaking of watching for signs of weather:

> You hypocrites! You know how to interpret the appearance of earth and sky (heaven), but why do you not know how to interpret the present time? (Luke 12:56 ESV)

Clouds, which are in the air, are vaporous water. Rain arises from the condensation of these clouds. Thus, there is air, heaven, or a firmament between the clouds and the earth. There is, in fact, a separation of the waters below from the waters above.

Augustine calls this account praiseworthy. Nevertheless, he will go on to adopt a different account. Later, he expresses some doubts about the air between the earth and clouds being called a firmament:[8]

---

[7] Augustine, *The Literal Meaning of Genesis. 1*, Book Two, Chapter 1.
[8] Augustine, Book 3, Chapter 1.

> For air is so near the heaven in which the luminaries shine it
> is also called heaven, but I do know whether it could likewise
> be called a firmament.

Augustine goes on to give his own account.[9] Augustine observes that water can rest above the air in clouds. He takes this to be because the water is divided into very small drops. He suggests that if the waters were divided into still smaller drops, then they could easily remain above even the higher heavens. He argues that every drop, no matter how small, can be divided into smaller drops. Consequently, if the drops were divided very finely, they could rest above all of the heavens.

Augustine records, without any explicit judgment, a scientific argument for the existence of the waters above the heavens.[10] The argument appeals to the movement of Saturn. Saturn was the farthest known planet; it takes about thirty years to circle the sun. Likewise, in the geocentric system, it took about thirty years to complete its cycle around the earth. The reason given for this slow movement was that Saturn traveled in a very large circle around the earth. However, at the same time, Saturn was rotating around the earth, along with the rest of the heavens, once per day. If Saturn was so far away, this means it must be moving at an incredible speed to make the daily rotation. This was thought to generate a lot of heat from the friction of the movement. Thus, we should be able to feel the heat of Saturn. However, Saturn does not give off any apparent heat. But the waters above could explains this by suggesting that Saturn is cooled off by these waters above the heavens, which are in an icy form.

Augustine concludes with a statement that we must believe

---

[9] Augustine, Book Two, Chapter 4, Section 8.
[10] Augustine, Book Two, Chapter 5, Section 9.

in the waters:[11]

> But whatever the nature of that water and whatever the manner of its being there, we must not doubt that it does exist in that place. The authority of Scripture in this matter is greater than all human ingenuity.

Augustine explained three different accounts of the waters. In *Against the Manichees* he understands them as spiritual waters, not part of the corporeal world. In *The Literal Meaning of Genesis*, he characterizes the idea that the waters above are clouds as a "praiseworthy theory." He ends by arguing that it is possible for waters to remain above the heavens by being in very small drops.

John Chrysostom

Chrysostom asks what is meant by the firmament.[12] He mentions the possibility of solid water or some sort of compressed air. However, he says that no sensible person would make a decision on it. Rather, we should accept what is told us: God created the firmament that carries the waters above it.

Severian of Gabala

Severian holds the firmament to be made out of water made solid.[13] Severian understands it to be called a firmament because it is firm relative to water. He attempts to support his contention that the firmament was made from solidified water by appealing to a Psalm in the Septuagint:

> By the word of the Lord the heavens were made firm, and by the breath of his mouth all their host (Psalm 33:6 - New English Translation of the Septuagint)

---

[11] Augustine, Book Two, Chapter 5, Section 9.

[12] Chrysostom, *Homilies on Genesis*, Homily 4.

[13] Severian, "Homilies on Creation and Fall," Homily Two.

The Hebrew simply indicates the heavens were made, with no indication that they were made firm. However, the Greek used a variant of *stereōma* meaning "made firm." The implication would be that the heavens were originally not firm and then were made firm. The heavens were thus not made of a material which was already firm, like a rock, but rather water which could be made firm.

Severian asks what the purpose of the waters above might be.[14] He references Psalm 148 to confirm the existence of these waters. God placed the water on the firmament, or crystal-like heaven, in order to keep it cool from the heat applied by the sun, moon, and stars. He compares this to a pot of water, where the water resists the fire.

Severian believes that the dew comes from the waters above. The rain comes from the clouds. However, there can be dew without clouds and thus Severian concludes that dew comes from the waters above the firmament. To support this idea, he references Biblical statements referring to the dew of heaven.

> May God give you of the dew of heaven and of the fatness of the earth and plenty of grain and wine. (Genesis 27:28 ESV)

Cosmas Indicopleustes

Cosmas mentions the waters several times in passages such as the following:[15]

> For the pagans who hold the theory of the sphere, if consistent with themselves, neither entertain such a hope, nor allow that there are waters above the heaven, nor are found to acknowledge that the heavenly bodies and the world will come

---

[14] Severian, Homily Two.

[15] Indicopleustes, *Christian Topography*, Book III.

to an end; but expect that the world in the state of corruption will continue for ever.

However, beyond insisting that those who accept pagan cosmology cannot believe in the waters, he says little about them. In his thinking, accepting the pagan cosmology of the sphere rules out the possibility of any waters even beyond the pagan's spheres.

John of Damascus

John asks why God placed water above the firmament.[16] He gives the reason that it was due to the intense heat of the aether. Without the water, the firmament would have been destroyed by the heat.

Bede

Bede says that there are waters above our heavens, the firmament.[17] He refers to Psalm 104, which speaks of God laying his beam in the waters:

> Bless the LORD, O my soul! O LORD my God, you are very great! You are clothed with splendor and majesty, covering yourself with light as with a garment, stretching out the heavens like a tent. He lays the beams of his chambers on the waters; he makes the clouds his chariot; he rides on the wings of the wind; (Psalm 104:1–3 ESV)

He therefore concludes that the firmament is between the upper and lower waters.

Bede considers the objection that the waters would not remain on top of the spherical heaven. In response, he appeals to Job:

---

[16] John of Damascus, "An Exposition of the Orthodox Faith," Book II, Chapter 9.
[17] Bede, "Commentary on Genesis."

> He binds up the waters in his thick clouds, and the cloud is
> not split open under them. (Job 26:8 ESV)

He then explains: If God can keep the waters in a cloud in a
vapor, he is certainly able to keep the waters above the heavens
in a solid sphere of ice. Furthermore, even if he wanted to keep
the liquid waters above the heavens, this is a small thing next
to fixing the earth on nothing, referring to another passage
from Job:

> He stretches out the north over the void and hangs the earth
> on nothing. (Job 26:7 ESV)

Bede points to God holding back the waters of the Red Sea and
the Jordan. If God can hold those waters back, clearly he can
hold the heavenly waters back. The Creator knows what type
of waters these are and their purposes. But as for us, there is
no room for doubt, these waters exist because Holy Scripture
says they do.

Remigius of Auxerre

Remigius gives a similar discussion as Bede.[18] He suggests that
the waters above are in place as ice or, by a miracle, in liquid
form. If God can suspend the waters of the Jordan or Red Sea
and hang the earth upon nothing, then he will have no difficulty keeping the waters aloft. He suggests that these waters
exist to temper the heat of the stars and to wash the earth on
the day of judgment.

---

[18] Remigius of Auxerre, "Exposition on Genesis," Chapter 1.

[19] Aquinas, *The Summa Theologiæ of St. Thomas Aquinas*, Part 1, Question 68,
Article 2.

Thomas Aquinas

Aquinas discusses the waters above.[19] Aquinas agrees with Augustine that the Scriptures have more authority than human intellect. As such, we must, as Christians, believe without doubt that these waters actually exist. Aquinas describes, but rejects, Origen's interpretation of the waters as spiritual substances. The water must be material.

Aquinas explains several different interpretations of the waters above. They might be another heaven or sphere above the starry heavens made of less dense water. There might be a heaven beyond the stars that is not made of water but described as waters due to being transparent. The waters might be a reference to formless matter. The waters might be clouds carrying water. Aquinas answers objections to these different views and does not explicitly take a position as to which he thinks is correct.

Aquinas takes particular note of an argument rooted in classical science that claimed that every element had its own natural place. There could not be two natural places for water, and thus there ought not to be a place for water on the earth as well as in the heavens. He resolves this by arguing that these were two different kinds or species of water and thus they could have two distinct natural places.[20]

If the heavens created on day two divided the waters below from the waters above, this would seem to imply that the waters below should be directly below the heavens.[21] However, it is evident that the waters do not reach up to the heavens, but rather there is air between them.

Aquinas says that "considered superficially," the creation account might lead one to conclude that the world was initially

---

[20] Aquinas, Part 1, Question 68, Article 3, Objection 1.
[21] Aquinas, Part 1, Question 68, Article 3, Objection 3.

an infinite volume of water. However, because "solid reasons" had demonstrated that this was not the case, "it cannot be held to be the sense of Holy Scripture." He argues that the initial world must have consisted not only of earth and water, but also air. Air is not explicitly mentioned because Moses wrote to an ignorant audience. His audience would not have understood that air was a substance and not just empty space. Instead, when Moses refers to waters, he also means to include the air, and thus there are waters up to the heavens. Air, Aquinas states, is described as water because air is transparent like water.

Martin Luther

Luther expresses his difficulty in understanding the waters above.[22] He identifies the waters below with the clouds that fly under the firmament. However, he states that Moses speaks, with the plainest possible words, of waters above the firmament. Luther notes that the theologians had added two spheres to the classical system: one that corresponded to the waters above, called aqueous, glacial, or crystalline, and the other, the supercelestial realm, called empyreal or fiery. Luther is suspicious of these additions, chiding Ambrose and Augustine for "childish thoughts" on this question relative to the elegant and prudent Greeks.

He mentions the idea that the waters above are responsible for the movements of the heavens. However, he thinks this idea is "puerile"; he would rather say that he does not understand than approve such a thought.

According to Luther, Moses makes it clear that there are waters above the heavens. However, we have no idea what these waters might be. He mentions one other reference to

---

[22] Luther, *Luther On The Creation: A Critical and Devotional Commentry on Genesis.*

these waters in the Apocrypha. He concludes that we simply do not know anything about this or related subjects such as the location of God, the departed saints, or His angels.

John Calvin

Calvin considers the division of the waters.[23] As he notes, it seems contrary to common sense to think that there might be waters above the heavens. Calvin argues that this text should only be understood as referring to things that we can see. This is the wrong place to look to find out what is in the distant reaches of space. He says, "he who would learn astronomy, and other recondite arts, let him go elsewhere." Consequently, whatever the text refers to, it must be something in our ordinary experience. Moses did not intend for us to hold by faith to the existence of distant waters.

Calvin argues that the waters above are merely a reference to the clouds. The clouds are suspended in the air. It is only by the providence of God that they are kept aloft and the waters do not fall upon them and flood the earth. Calvin quotes both references to the waters above the heavens in the Psalms as David praising God for his miracle of keeping the waters aloft.

Conclusions

Historical Christians

Many authors note the difficulty in understanding the references to the waters above. A frequent theme is that even though we do not understand what the text is referring to, we are obligated by the authority of Scripture to believe in these waters. Some such as Theophilus and Basil understand these

---

[23] Calvin, *Calvin's Commentary on the Bible*.

waters as being located in the upper regions of heavens and providing a source of rain and dew. Later authors typically understood the waters as being beyond the heavens and took the firmament as some sort of impassable barrier preventing the waters from falling to earth. There are references to an interpretation in which the waters above are clouds, but only Calvin explicitly endorses this as the correct interpretation.

A variety of explanations have been given for why these waters do not fall to earth. Some appealed to the firmament. Others argued that the waters were of a different nature than the waters on earth. Still others simply considered it a miraculous act of God to keep the waters in place.

The firmament itself is typically understood to be a barrier to the waters. However, these authors often indicate that it is firm and solid only in a relative sense. They did not think it was as firm and solid as the earth. The firmament was strong and unyielding but not quite a solid and hard body in the same sense as the earth.

Modern Science

Tycho Brahe was the last great naked eye astronomer. He is best known for the Tychonic system, which combined the classical geocentric system with Copernicus's then-new heliocentric system. He also is known for determining that comets came from the heavens into the earthly spheres. This demonstrated that the heavens were not a hard and impenetrable barrier.

Until the 1980s, it was generally accepted that classical science held to hard and impenetrable barriers. However, more recent scholarship has revised this opinion. Despite Brahe's believing that he was overturning a widespread traditional view, the idea does not seem to have wide currency much prior to this point.

The classical scientific argument that there could not be waters beyond the heavens relied on the classical conception of gravity. If all waters anywhere in the physical cosmos fall towards the earth, it is difficult to see how there could be waters far above in the heavens. However, under a modern scientific conception of gravity, this is no longer a difficulty.

Today, we understand that there is not only water in the clouds but actually in the clear parts of the sky as well. There is a large amount of water vapor in the atmosphere, and clouds are formed when this vapor condenses into tiny water droplets. Beyond that, there is a large amount of water spread throughout the universe.

Nevertheless, modern cosmology has not discovered a massive body of water surrounding the heavens. On the other hand, it is difficult to rule out the possibility that there might be such a body outside of the observable universe. Most physicists and astronomers would strongly reject the idea, but we cannot prove what exists or does not exist outside of the observable universe. Some young-earth creationists have argued for the existence of a cosmic shell of water beyond the observable universe. Nevertheless, for many Christians, maintaining the existence of this shell would be difficult.

Evaluation of the Response

Historically, most Christians thought there was some sort of body of water in or beyond the heavens, held up by a firmament. There were exceptions, such as Origen, who interpreted it as a reference to spiritual beings, or Calvin, who thought it was a reference to clouds. Some other authors, such as Augustine and Aquinas, also mention the interpretation of the waters above as clouds but do not endorse it. However, neither the idea that there is a vast body of water in or beyond the heavens or the idea of solid firmament has fared well in the light of modern science. Where did historical Christians go wrong?

Part of the problem is the Septuagint which translated the account of the second day in a way that strongly suggested a barrier holding up waters beyond the heavens. Most notably, the word *raqia*, of uncertain meaning, was translated in a way that referred to something firm and solid. Moreover, the other language in the Septuagint creation account also reinforces this idea. This resulted in translations that seemed to very clearly teach the existence of waters above and beyond the heavens.

Another reason was classical cosmology and gravity. Within that system, the idea of a firmament and the waters above seem to fit together nicely. The waters give the firmament a purpose, and the firmament explains why the waters do not fall. Since the heavens were already conceived of as a series of concentric spheres, incorporating the firmament and waters seemed very obvious.

The propensity of historical Christians to insist on these waters and the barrier reflected a commitment to the authority of Scripture. Even though the nature of the heavens was a scientific rather than a philosophical issue, they insisted on following what they took the Scripture to be teaching. They were even willing to go against the science of their day on this issue. Ultimately, we can see that they were incorrect, but their insistence still shows the commitment they had to the text.

The identity of the firmament and the waters above remain an issue today. Today, Bible believing Christians often either try to take the waters above as reference to the clouds, or postulate bodies of water beyond the stars, or try to dismiss it as phenomenological language. However, all of these explanations have significant issues. In my view, we should return the source and understand the second day as a division of the waters into the skies above and the land and seas below.

Chapter 10

Is There Water Under the Earth?

The Scientific Question

What is under the earth? Before answering that question, we should clarify what we mean by earth. Today, the word "earth" primarily refers to our planet as a whole. However, historically, the English word earth, as well as its Greek, Latin, and Hebrew equivalents, referred instead to the ground. Even today, the word earth can still refer to dirt instead of the planet as a whole. When the ancients asked what was under and supporting the earth, they were not asking what was under the planet as a whole, but what was under the ground that we walk on.

An obvious answer would be that the ground rests on some sort of solid foundation. After all, what else could hold up the earth? The nature of such a foundation might be vague or unclear in the ancient mind. It could be conceived as something like a layer of solid rock holding up the ground. Whatever the nature of the foundation, it holds up the ground as we know it.

However, classical science rejected the existence of such a foundation. For classical science, earth was one of only four elements. None of the other elements—air, fire, or water—could hold up the earth. Earth, they concluded, was at the bottom of the universe; there was nothing underneath it. They considered the entire sphere making up our planet to mainly consist of the element earth, and this eventually led to us calling our planet the earth.

However, earlier philosophers, such as Thales of Miletus, the first Greek natural philosopher, had proposed that the earth floated on water. He thought of the earth as a floating island on top of the ocean. An obvious difficulty is that dirt or earth does not float. While some things, like logs or ships, do float on the water, dirt or earth just sinks. It went quite against the observed nature of things for the earth to float on the waters.

The Biblical Discussion

Some verses suggest that the earth has a foundation:

> Where were you when I laid the foundation of the earth? Tell me, if you have understanding. (Job 38:4 ESV)

Others suggest that the earth is on top of the seas:

> A Psalm of David. The earth is the LORD's and the fullness thereof, the world and those who dwell therein, for he has founded it upon the seas and established it upon the rivers. (Psalm 24:1–2 ESV)

Still others suggest that the earth is set on pillars:

> He raises up the poor from the dust; he lifts the needy from the ash heap to make them sit with princes and inherit a seat of honor. For the pillars of the earth are the LORD's, and on them he has set the world. (1 Samuel 2:8 ESV)

Another verse suggests that the earth is not hanging on anything:

He stretches out the north over the void and hangs the earth on nothing. (Job 26:7 ESV)

As such, Biblical texts could be used to argue that the earth rests on a solid foundation, floats on the oceans, is set on pillars, or hangs on nothing at all. What are we to make of all this? An important observation is that almost all of these passages are found in poetic contexts. This means that they have to be interpreted with a certain degree of poetic license. Nevertheless, poetic license is not unlimited.

There is one interpretation that can make sense of these texts while invoking minimal poetic license. The earth, i.e., the ground we walk on, is supported by a further solid foundation. Today, we would think of this foundation as being the bedrock; however, the Biblical conception is much vaguer. The Biblical text refers to a foundation of the earth without any clear indications of the properties of this foundation.

There are several references to either the foundation or foundations of the earth. In the English Standard Version, the phrase "foundation of the earth" or "foundations of the earth" appears twelve times. Other passages refer to God founding the earth (Psalm 24:2, Psalm 78:69, Psalm 89:11, Psalm 102:25, Proverbs 3:19, Isaiah 48:13, Isaiah 51:13,16, Zechariah 12:1). In all these cases, the word being translated is *yasad*, which usually refers to laying down a foundation, or *yesod*, a related noun referring to a base or foundation. Crucially, there are far more verses using this language of a foundation than can be appealed to in support of any other interpretation.

The earth is described as being above the waters and the waters as being beneath the earth in several passages. For example:

> You shall not make for yourself a carved image, or any like-
> ness of anything that is in heaven above, or that is in the earth
> beneath, or that is in the water under the earth. (Exodus 20:4
> ESV)

> A Psalm of David. The earth is the LORD's and the fullness
> thereof, the world and those who dwell therein, for he has
> founded it upon the seas and established it upon the rivers.
> (Psalm 24:1–2 ESV)

However, as we have seen previously, Hebrew prepositions
generally have a wider range of meaning than English prepositions. The word translated as above, *al,* is usually translated
as by or beside when referring to water:

> Now the daughter of Pharaoh came down to bathe at the
> river, while her young women walked beside (*al*) the river.
> She saw the basket among the reeds and sent her servant
> woman, and she took it. (Exodus 2:5 ESV)

The young women are not walking on or above the river.
The word translated as beneath is *tachath* and is used in the
following verse to refer to the foot of the mountains:

> And Moses wrote down all the words of the LORD. He rose
> early in the morning and built an altar at the foot (*tachath*)
> of the mountain, and twelve pillars, according to the twelve
> tribes of Israel. (Exodus 24:4 ESV)

Moses was not building an altar beneath the mountain; he
was doing so at the bottom or foot of the mountain. The Bible
describes the earth as being above the water in the sense that
the land is at a higher elevation. It is not saying the earth is
floating or resting on the waters.

Hannah's prayer speaks of the world being set on pillars.
However, she does not use either of the ordinary words for

pillar, but a word of unclear meaning. Its only other use is later in 1 Samuel:

> The one crag (pillar) rose on the north in front of Michmash, and the other on the south in front of Geba. (1 Samuel 14:5 ESV)

This verse is referring to a rock face, not a pillar. The semantic range of this rare word is unclear, but plausibly this is simply a reference to the foundation of the earth that we read about in other places.

A couple of other passages refer to the pillars of the earth being shaken:

> who shakes the earth out of its place, and its pillars tremble; (Job 9:6 ESV)

> When the earth totters, and all its inhabitants, it is I who keep steady its pillars. (Psalm 75:3 ESV)

Another similar verse refers to the pillars of heaven:

> The pillars of heaven tremble and are astounded at his rebuke. (Job 26:11 ESV)

I would suggest that these verses are making poetic reference to mountains being shaken by earthquakes. Furthermore, there is nothing in them that actually indicates that the earth is resting on these pillars. They are instead merely associated, in some way, with the earth and the heavens.

Job contains a statement that God made the earth hang on nothing:

> He stretches out the north over the void and hangs the earth on nothing. (Job 26:7 ESV)

However, a more mundane explanation is that the verse simply gives an account of creation. While the meaning of the reference to the north is debated, a common suggestion is that it refers to the heavens. In the northern hemisphere, all of the stars circle around the north pole. North is thus the direction of heaven, and by north, the text is referring to the heavens. This fits with the common Biblical motif that God stretched out the heavens.

The void referenced in the verse is *tohuw*, the same word that is commonly translated as "formless" in Genesis 1:2. This suggests that the verse is speaking of God's creation of the heavens from a formless state. The parallelism suggests that something similar is intended in the second part of the verse. The earth, like the heavens, were created from the nothingness that persisted at creation. Consequently, it is plausible that the verse poetically describes the creation of the world and has no intention to make any statement about the earth's support.

The creation and flood accounts are difficult to reconcile with the idea that the earth is floating on the ocean. The earth is described as appearing when waters were pulled away from its surface and being flooded when a great excess of water was added. If the earth floated, it would make more sense to describe the earth as rising out of the waters and sinking into them. Indeed, no amount of additional water added to the ocean would make the earth flood because the earth would simply rise with the rising water.

Altogether, the idea that the ground is supported by a foundation is the most plausible interpretation of the text. This accords with the most common Biblical language, referring to a foundation for the earth. A closer examination of the apparent support for other ideas reveals that the original text does not support them very strongly. The idea of a floating earth, in particular, conflicts with the creation and flood accounts. As such, the earth resting on a foundation seems the most plausible interpretation.

Historical Christians

Athanasius of Alexandria

Athanasius argues that the order found in the world points to God.[1] In particular, he points to examples of phenomena that defied understanding by classical science. Of interest to us is his second example: The earth is very heavy and thus ought to sink, but instead it remains fixed on top of the ocean. The waters, despite their nature, support the earth.

Athanasius thinks that the earth is fixed on top of and supported by the ocean. It is not sitting on some sort of foundation or pillar. This, he argues, is a testament to the power of God. However, he does not invoke Scripture or theology to show that the earth is floating on the waters. Rather, he simply assumes it, as though it were a commonly accepted idea in his day.

Basil of Caesarea

Basil discusses the earth's support.[2] He points out that the earth cannot be supported by air. The air would slip away in all directions when the weight of the earth pressed upon it. It could also not be supported by water because it would sink. Even if it were supported by water, that would merely leave the question of what supported the water.

Basil argues that positing a heavy body to hold up the earth would also not work. While, unlike water or air, the heavy body could readily hold up the earth, it would itself need to be supported. Even if a support for the earth were postulated, that support would itself have to be supported by another foundation. Furthermore, each foundation would have to be stronger than the last in order to support the previous foundation.

---

[1] Athanasius of Alexandria, "Against the Heathen," Chapter 36.
[2] Basil of Caesarea, "Hexaemeron," Homily I, Sections 8-9.

There would be no end to the infinite regress of the stronger foundations required to support the earth.

Basil pointed out the difficulties of various possible supports for the earth. However, he does not argue for an alternative. Instead, he argued that attempting to understand what supports the earth is "investigating the incomprehensible." Basil refers to the Book of Job:

> Where were you when I laid the foundation of the earth? Tell me, if you have understanding. Who determined its measurements—surely you know! Or who stretched the line upon it? On what were its bases sunk, or who laid its cornerstone, (Job 38:4–6 ESV)

Basil's argument is that if we are arrogant enough to think that we understand what God made to support the earth, then we run the risk of being reproached by God just as Job was.

Basil refers to Psalm 75, which mentions that God keeps the pillars of the earth steady. He briefly indicates that the pillars are a metaphor for the power of God, which sustains the earth. This is not an entirely satisfying answer because the pillars are being kept steady, and what this means if the pillars are a metaphor for God's power is unclear.

Basil also refers to Psalm 24, which speaks of the earth's being founded on the seas. He takes the reference to state not that the earth is above the waters but that "water is spread all around the earth." However, he does not elaborate on what he means by this or how it can be a reasonable interpretation.

Basil briefly argues that the possibility that the earth is suspended on nothing is even harder to explain. It has a heavier nature than the other elements and thus ought to fall down. Thus, it is even more incomprehensible to think that the earth is supported by nothing at all than it is to think it is supported by all the other elements.

Despite having criticized these ideas, Basil allows that the earth may rest upon itself or on the waters, but insists that the important thing is to believe that it is sustained by the Creator's powers. If we are asked what supports the earth, Basil says that we should answer that it is held up by the hand of God. The two ideas he endorses are those of classical science: the earth resting on itself or a floating earth, suggesting that they were both popular theories in his time.

Augustine of Hippo

Augustine mentions this issue while arguing that earth is heavier than water.[3] Here he warns against attempting to argue that the earth is lighter than water just because the psalmist indicates that the earth was established above water. In particular, we should not use this verse to dispute the findings of people who have studied the elements and determined that the earth is heavier. If we do, they will scorn the books of Scripture rather than abandon the knowledge they have by unassailable arguments or experience.

Augustine offers an allegorical understanding of this passage. Heaven is the understanding of the truth. Earth is the simple faith of children. But this simple faith is made firm by baptism. Hence, the earth (the simple faith) is established by water (baptism).

However, Augustine is not willing to insist on the allegorical interpretation. Instead, he gives a literal interpretation as well, in case anyone insists on a literal interpretation. He suggests that the earth above water may refer to land that happens to jet out over the water or the roofs of underground caverns with water in them.

---

[3] Augustine, *The Literal Meaning of Genesis. 1*, Book 2, Chapter 1, Section 4.

Augustine's interpretation is rather awkward. It is pretty implausible to think that the text speaks of promontories or cavern roofs. Part of Augustine's difficulty is that while the Septuagint and Hebrew texts use a preposition with considerable range, his Latin translation used "super" which means above and has much less semantic range.

John Chrysostom

Chrysostom draws attention to the fact that it is not the nature of water to support the earth.[4] Instead, it is the nature of the earth to support the waters. Even a light pebble will fall through the waters. Nevertheless, Chrysostom thinks that the whole earth is floating on the waters. According to Chrysostom, God, by his divine power, is keeping the earth above the waters.

Chrysostom points to Biblical passages in order to support the idea that the earth floats on the waters. In fact, he suggests that this idea comes from "the prophet." The "prophet," in this case, is David and another psalmist.

> And whence does this appear, that the earth is borne upon the waters? The prophet declares this when he says, "He has founded it upon the seas, and prepared it upon the floods." And again: "To him who has founded the earth upon the waters."

Chrysostom makes the point that water cannot even support a pebble. How could it possibly support the earth with its mountains, hills, cities, men, and animals without becoming submerged?

Chrysostom points out that the earth ought to be broken apart by the waters underneath it. In water, earth turns to mud, wood goes rotten, and iron goes soft. He points to an example

---

[4] Chrysostom, "Homilies on the Statues," Hoomily 9.

of run-away slaves who soak shackles and chains in order to make them easier to break. How is it that the earth has not been destroyed by this constant contact with water?

Chrysostom invokes three seemingly contradictory passages about the support of the earth:

> He stretches out the north over the void and hangs the earth on nothing. (Job 26:7 ESV)

> In his hand are the depths of the earth; the heights of the mountains are his also. (Psalm 95:4 ESV)

> for he has founded it upon the seas and established it upon the rivers. (Psalm 24:2 ESV)

However, Chrysostom argues that these passages all mean the same thing. He thinks that the earth floats on the water, which easily explains the psalmist speaking of the earth being founded on the seas. But since the water is not a solid support, it is effectively the same as if it were supported by nothing, which explains the reference in Job to the earth's hanging on nothing. The psalmist also speaks of the hand of God's holding up the earth, but this refers to God's power's holding and supporting the earth, not a literal hand physically holding up the earth.

Chrysostom argues that the Scriptures teach that the earth is fixed on the waters. It cannot be quite said to float on the waters because it is held above the waters by the power of God rather than floating naturally. He does not show any awareness of the classical theory of gravity or alternative accounts of the support of the earth. Unlike his rejection of the spherical heavens, he does not make any explicit attack on an opposing view. This suggests that he takes the earth's floating on the ocean as the standard view of his day.

John of Damascus

John of Damascus speaks of the earth and indicates that we do not know what supports it.[5] Some believe it is the water and cite the psalmist to support this claim. Others suggest it hangs in the air. Others cite Job, declaring that it hangs on nothing. David refers to the pillars of the earth, meaning the force that sustains the earth. Ultimately, John indicates that whatever we think about the support of the earth, we must admit that all things are sustained and preserved by divine power.

Conclusions

The Christian Interaction

We find that some of the earlier authors, such as Athanasius, Chrysostom, and Basil, think or find it plausible that the earth is floating on the ocean. None of these authors give any indication that this was a controversial view in their day. Only Chrysostom tries to defend it as the teaching of Scripture. John of Damascus, who lived much later, includes it in a list of possible understandings of the support of the earth.

Augustine rejects the idea. He appears to be more influenced by the ideas of later classical science and views the idea of a floating earth as a clearly flawed scientific theory. He very strongly urges Christians not to use the Scriptures to defend that view. However, he was not able to provide a compelling alternative exegesis.

---

[5] John of Damascus, "An Exposition of the Orthodox Faith," Book II, Chapter 10.

Modern Science

Earth does not float on the oceans or through the air. The ground we walk on is ultimately supported by bedrock, the mantle, and the core. But if we consider the whole mass of what we now call the earth, the classical science theory that the earth is supported on itself is essentially correct.

Evaluation of the Response

Athanasius and Chrysostom defend the idea that the earth is floating on the waters. This has turned out to be incorrect. Why did they come to this errant view? Partially, they seem to have taken it as the standard view of their day. Athanasius, in particular, does not appeal to any Scripture to support it. Chrysostom does, but it seems, at least in part, to be a matter of reading their scientific theory into the text. The texts they appeal to are, at best, ambiguous supports for their theories.

The idea that the Bible taught that the earth was floating in the oceans saw little support and was rejected. We can take encouragement from this. The Church was not drawn into flawed cosmologies by errant interpretations of the Biblical text. Exegesis suggesting that the earth floated on waters or was standing on pillars was considered and rejected by the Christian Church.

Nevertheless, we see that historical Christians accepted the authority of the Biblical text. None of these authors rejected the Biblical texts or dismissed the possibility that they might speak of these scientific matters. Those who rejected the claims attributed to the text did so by arguing that the text meant something else.

We also see that the solution was going back to the source. Errors arose from incorrect translations and errant but common scientific ideas. When we go back to the source, we find that the Bible simply did not teach those ideas and that they were read into the text.

Chapter 11

Why Isn't the Earth Covered With Water?

The Scientific Question

According to classical science, our world consists of four elements: earth, water, air, and fire. Furthermore, these elements fit into a series of concentric spheres around the center of the world according to their weight. The heaviest element, earth, was at the center. Water, making up the oceans, rested on top of the earth. There was a layer of air above this. Above the air was a fiery realm, sometimes considered to contain the heavenly bodies, which were thought to be made of fire.

However, there was something not quite right about this picture. There are oceans above some parts of the earth, but other parts of the earth are clearly above the ocean. The layers are not as neatly sorted as this depiction suggests. Why is the earth poking up beyond the ocean? If the world is best understood as a series of concentric spheres, why is the earth not completely submerged?

One possible answer is that the earth is not perfectly spherical. It contains high mountains and deep holes. We know these deep holes as the ocean basins. This makes it possible that parts of the earth were above the waters while other parts held the oceans. Since the element earth was hard and solid, it made sense that it could maintain this non-spherical shape.

However, rivers are seen to constantly flow into the oceans. Why are these basins not filling up and overflowing the land? Part of the answer is that these rivers are fed by rain that comes from water that evaporates from the ocean. Hence, the water in rivers is simply returning to the ocean whence it came. However, it seemed to those in the ancient world that there was far more water in the rivers than could be explained by rain. It was thought that much of the water came from springs, which ancient people generally thought came from the sea through underground passages.

An alternative possibility was that the oceans were actually at a higher elevation than the land, but something was preventing them from overflowing the earth. Instead of the oceans sitting in basins carved out of the land, a section of the ocean was "carved out" to allow there to be dry land. This sounds very strange to modern ears. Nevertheless, it was the dominant understanding at the end of the Middle Ages. It does not appear to have come from the classical world but was developed during the rediscovery of classical science in the later Middle Ages.

Part of the reason this idea gained currency was that, in some situations, it can appear to be true. If you stand on the seashore and look towards the ocean, you can sometimes get the impression that the water is rising as it moves away from the shore. Likewise, sailors moving towards the shore perceive the tops of mountains as being at the same level as themselves.

Another reason for this idea was to explain the distribution of land. As far as the ancients knew, there was one large land mass (the combination of Europe, Asia, and Africa). Why was all the land concentrated on one part of the globe? This could be explained by the idea that the oceans were generally higher than the land, but that there was something special about the particular area where people lived that caused it to be free of water.

The Biblical Discussion

The Genesis 1 creation account describes the gathering of the waters and the appearance of the dry land:

> And God said, "Let the waters under the heavens be gathered together into one place, and let the dry land appear." And it was so. God called the dry land Earth, and the waters that were gathered together he called Seas. And God saw that it was good. (Genesis 1:9–10 ESV)

What exactly does it mean to gather together the waters? The implication seems to be that there were waters covering the earth that were removed, revealing it. But where did all of that water go? What happened to the mass of water that would have been required to cover the whole earth?

Psalm 104 gives a description of the creation of the land and seas:

> He set the earth on its foundations, so that it should never be moved. You covered it with the deep as with a garment; the waters stood above the mountains. At your rebuke they fled; at the sound of your thunder they took to flight. The mountains rose, the valleys sank down to the place that you appointed for them. You set a boundary that they may not pass, so that they might not again cover the earth. (Psalm 104:5–9 ESV)

In this depiction, the earth is covered with water. God rebukes the waters, causing them to flee. In this translation, the mountains rise while the valleys sink. In some other translations, the waters are said to flow over the mountains and into the valleys.

In either case, the text seems to suggest that waters flow into valleys, which become ocean basins.

The whole account is obviously poetic. We should perhaps not take it as a precise description of the process of creating the sea and dry land. Nevertheless, the imagery seems to assume that seas are resting in basins.

A verse from the Psalms refers to the waters being gathered into a heap:

> By the word of the LORD the heavens were made, and by the breath of his mouth all their host. He gathers the waters of the sea as a heap; he puts the deeps in storehouses. (Psalm 33:6–7 ESV)

One could argue that waters could not be described as being gathered into a heap unless they piled upwards, thus implying that the waters were above the level of the land. Furthermore, almost all other uses of the Hebrew word translated as heap, נֵד (ned), refer to the miraculous piling up of waters during the crossings of the Jordan River or the Red Sea. On the other hand, this is a poetic passage. We do not think that God literally placed the water into a storehouse, and neither should we think the water was literally placed in heaps.

Certain verses speak of God's erecting barriers that the sea cannot cross.

> You set a boundary that they may not pass, so that they might not again cover the earth. (Psalm 104:9 ESV)

> when he assigned to the sea its limit, so that the waters might not transgress his command, when he marked out the foundations of the earth, (Proverbs 8:29 ESV)

> Do you not fear me? declares the LORD. Do you not tremble before me? I placed the sand as the boundary for the sea, a perpetual barrier that it cannot pass; though the waves toss,

they cannot prevail; though they roar, they cannot pass over
it. (Jeremiah 5:22 ESV)

These verses refer to barriers or boundaries for the sea. Jeremiah explicitly identifies sand as this boundary. These verses could be taken as referring to some sort of physical barrier that prevents the waters from rising up. However, the words used to describe it are *gebul*, translated as boundary, which is used to describe the borders of a territory, and *choq*, translated as limit or barrier, which refers to a legal statute. As such, these verses do not suggest a physical barrier but rather that God has set the place for the sea.

We saw in the previous chapter that the Bible describes the waters as being below the earth:

You shall not make for yourself a carved image, or any likeness of anything that is in heaven above, or that is in the earth beneath, or that is in the water under the earth. (Exodus 20:4 ESV)

However, if the seas are above the earth, this is problematic. As we argued, the text does not demand that the waters be directly below the earth. However, the semantic range could not allow the waters' actually being above the earth.

Furthermore, the flood is described as water's raining upon and increasing on the earth:

The flood continued forty days on the earth. The waters increased and bore up the ark, and it rose high above the earth. (Genesis 7:17 ESV)

This makes sense if the waters are below the earth and rise up to overflow it. It makes less sense if the waters were held at bay by divine force and then released to cover the earth.

There are some passages that can be construed as referring to the seas' being above the land. However, a closer inspection of the verses does not support that reading. Furthermore, other verses seem very awkward to interpret in a way that allows the waters to be considered to be above the land. Indeed, the better case is that Scripture assumes that the oceans rest in basins.

Historical Christians

Tertullian

Tertullian takes the waters' being gathered together as withdrawing into their hollow abysses, the ocean basins.[1]

John Chrysostom

Chrysostom speaks of the ferocious nature of the sea.[2] Yet he notes that it is walled in by feeble sand. God did not make the sea gentle, lest we think the sea does not overflow the land because of its gentle nature. Nor did he affix the land with a strong barrier to keep the sea away, lest we think that barrier prevents the sea from overflowing the land. It is the power of God to restrain the sea rather than any natural phenomenon. In support, he cites Jeremiah, describing the sand as the barrier to the sea.

Ephrem the Syrian

Ephrem understands the waters as flowing into basins.[3] He suggests three possible ways this may have happened. Firstly, God may have lowered the land in some places, creating the basins. Secondly, the waters may have "swallowed each other," thereby reducing the space required for them. Thirdly, a shak-

---

[1] Tertullian, "Against Hermogenes," Chapter 29.
[2] Chrysostom, "Homilies on the Statues," Homily 9.
[3] Ephrem the Syrian, *Commentary on Genesis*.

ing of the place of the sea might have made it a great basin that could collect the seas.

Basil of Caesarea

Basil comments on this issue while discussing the gathering of the waters.[4] When God ordered the waters to be gathered together, they flowed into the ocean basins. Basil explains that the basins were created when the waters needed to unite in a single mass. God created a vast space "all of a sudden" and the waters flowed into it. Once there, they were obliged by God's command to stay there. Without God's command, they might rise up and cover all the land. We see the strong waves break up on the sand. This is despite the fact that sand is the weakest substance. Basil indicates that the sand actually curbs the violence.

Basil claims Egypt is lower than the Red Sea, and without God's limitations, the Red Sea would overflow Egypt. He says that we know that the Red Sea is higher than Egypt due to failed attempts to build a canal connecting the Red Sea to the Mediterranean Sea. It was abandoned because it was found that it would cause Egypt to become flooded.

Actually, the canal was built. It did not connect the Red Sea directly to the Mediterranean Sea, as does the modern-day Suez Canal. Instead, it connected the Red Sea to the Nile delta. Parts of Egypt are indeed below sea level, but not the whole country. The ancient canal would have passed through what is now the Bitter Lakes. Before the Suez Canal was built, this was a below-sea-level valley. Due to the canal connecting this area to both the Mediterranean and Red Seas, it is now a pair of salt lakes. This is plausibly the historical basis for the idea that the

---

[4] Basil of Caesarea, "Hexaemeron," Homily IV, Chapter 3.

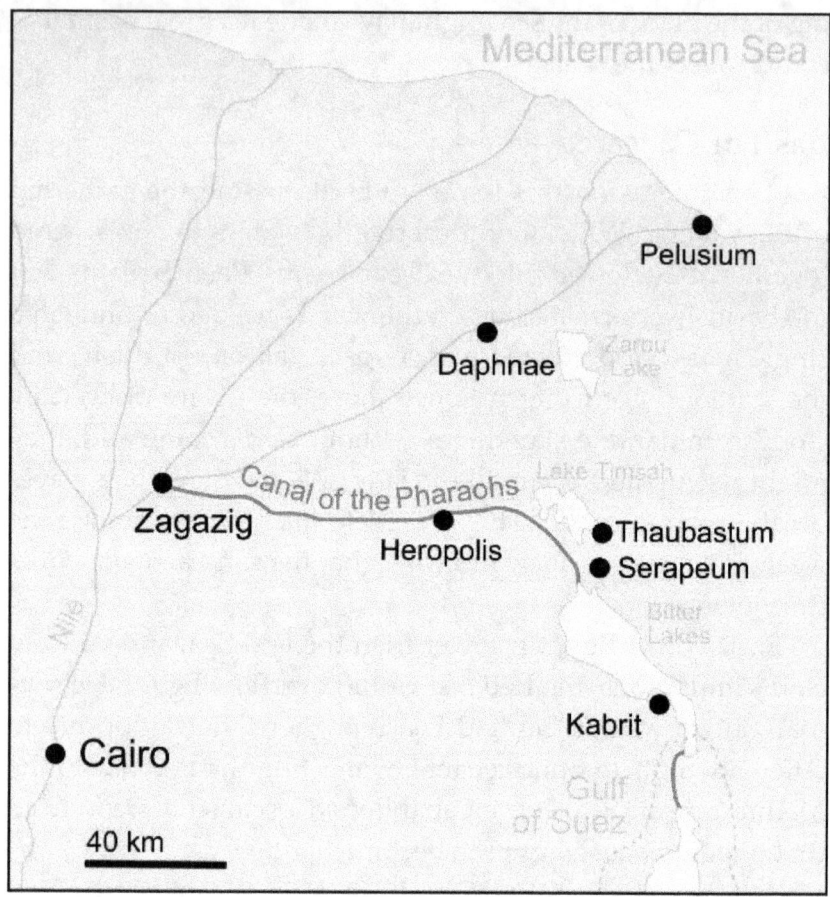

Figure 11.1: The path of the canals

project had to be abandoned due to the sea's being higher than the land.

Ambrose of Milan

Ambrose says that when God commanded the waters to be gathered together, this made the waters flow into the ocean basins and then stay *there*.[5] God has set up a boundary to

---

[5] Ambrose of Milan, "Hexameron," Book 3, Chapter 2, Section 10.

circumscribe the seas. He cites Job, with God describing his holding back of the waters:

> Or who shut in the sea with doors when it burst out from the womb, when I made clouds its garment and thick darkness its swaddling band, and prescribed limits for it and set bars and doors, and said, 'Thus far shall you come, and no farther, and here shall your proud waves be stayed'? (Job 38:8–11 ESV)

He points out that we often see the waves of the sea crashing into the shore. But the sea is held in check by the most unstable of all things, sand. Ambrose, like Basil, appeals to the idea that Egypt is below the Red Sea.

Ambrose asks how a single body of water could drain the whole land. It does not make sense to say that the earth appeared unless it was covered everywhere in water. Furthermore, if the flood of Noah hid even the mountains, surely so did the waters in the initial creation. But where did this over-abundant supply of water go?

Ambrose does not give a response, instead saying that he passes over it as a mystery. The Scripture does not give any clear testimony on the subject. Ambrose makes a somewhat vague reference to God's "enlarging space." This is probably a reference to the idea that ocean basins were enlarged in order to hold the waters. He says that God could have extended the low-lying regions, extending the basins, appealing to Isaiah for support:

> I will go before you and level the exalted places, I will break
> in pieces the doors of bronze and cut through the bars of iron,
> (Isaiah 45:2 ESV)

He also suggested that the force of water could make the ocean bed deeper, suggesting something like the flowing waters cutting into the land to make it deeper.

Augustine of Hippo

In *The Literal Meaning of Genesis*,[6] Augustine argues that creation was instantaneous. God did not create the world over the course of any time at all. As such, the earth was never covered with water. As part of the argument for his position, Augustine discusses the problem of explaining where the waters were gathered to.[7]

There could not be any bare portion of the earth because that would mean that there was already dry land. But if all the land was covered, where could the waters be gathered to? Augustine forcefully rejects the idea that the waters could be piled up above sea level. The ocean is clearly level. He admits that there are storms that deviate from this, but these are only temporary. There are tides, but this means merely that the waters are on other shores.

Augustine suggests that the water might be in a rarefied state, more like a cloud. It became dense and thus took up less room, allowing the land to appear. Or perhaps the land settled in vast areas, creating ocean basins for the water to flow into. Augustine neither endorses nor critiques these positions. We would expect that Augustine would attempt to critique both possibilities in order to provide support for his instantaneous creation interpretation. The fact that he doesn't suggests that he does not see a problem with them.

Bede

Bede brings up the question of where the waters were gathered to.[8] He suggests that the earth itself subsided to make room for

---

[6] Augustine, *The Literal Meaning of Genesis. 1.*
[7] Augustine, Book One, Chapter 12, Section 26.
[8] Bede, "Commentary on Genesis."

the waters. He also suggests that the waters were made more attenuated, i.e., less dense.

John of Damascus

John repeats the same idea of hollow places' being made in the earth that gathered the waters.[9] He adds the additional detail that this is how mountains are formed. He does not elaborate, but it may mean that God raises the land to form mountains while sinking the valleys to form seas.

Remigius of Auxerre

Remigius asks where so much water that could cover the entire earth could have gone.[10] He suggests that the earth sank down in certain places to receive the water. He also suggests that the water used to be much thinner and thus takes up less space now that it is denser.

Dante Alighiere

In 1320, Dante Alighiere (best known as the poet who wrote the *Divine Comedy*) delivered an address arguing against the idea that the waters were above the land, known as *A Question of the Water and of the Land*.[11] Dante does not invoke either Biblical passages or theology at any point. He discusses the same arguments discussed in this section as well as others, both scientific and philosophical.

---

[9] John of Damascus, "An Exposition of the Orthodox Faith," Book II, Chapter 9.
[10] Remigius of Auxerre, "Exposition on Genesis."
[11] Dante Alighieri, *A Question of the Water and of the Land*.
[12] Luther, *Luther On The Creation: A Critical and Devotional Commentry on Genesis*, 1:Chapter 1.

Martin Luther

Martin Luther refers to the tempestuous seas.[12] He says that the seas are higher than the land and thus would naturally overflow it. However, the spot of dry land where humans lived has an insuperable barrier, keeping back the waters. He cites Job 38 and Psalm 104 as support for this idea.

Luther compares the waters' being above the land to the children of Israel's crossing the Red Sea. In both cases, the waters stood higher than the land. God keeps the waters at bay. However, during the flood, these waters burst out and covered the earth. God also sometimes allows small islands to perish to remind us that he controls the waters.

Luther explains that Moses describes the fountains of the great deep as being rent open with violence. However, he says the Scripture is adapting itself to our understanding. It means rather that God ceased to restrain the water, letting them instead break forth unchecked according to their nature.

John Calvin

Calvin interprets the gathering together of the waters as their being placed above the sea.[13] Calvin discusses this issue in numerous places in his commentaries. We will only look at a few examples. He comments on Psalm 33:7:

> He gathers the waters of the sea as a heap; he puts the deeps in storehouses. (Psalm 33:7 ESV)

Calvin takes the reference to the seas' being piled in a heap to indicate that they are indeed piled upwards. The earth is lower and would, by nature, be flooded. God has enclosed the seas with invisible barriers that prevent them from overflowing the land.

---

[13] Calvin, *Calvin's Commentary on the Bible*, Commentary on Genesis 1:9.

He also comments on Psalm 24:2:

> for he has founded it upon the seas and established it upon the rivers. (Psalm 24:2 ESV)

Calvin argues that David is not speaking philosophically but rather using popular language. He adapts himself to the capacity of the unlearned, who wouldn't understand that the earth is below the seas. However, even this way of speaking is "not without reason." He means that there is some sense in which this is true. It is not entirely clear in what sense he means.

He also comments on Psalm 104:

> You covered it with the deep as with a garment; the waters stood above the mountains. At your rebuke they fled; at the sound of your thunder they took to flight. The mountains rose, the valleys sank down to the place that you appointed for them. You set a boundary that they may not pass, so that they might not again cover the earth. (Psalm 104:6–9 ESV)

Calvin indicates that there are two ways to take the reference to the earth's being covered with the deep as a garment. It might refer to the earth's being covered with waters during creation, or it might refer to the sea's now covering the earth. Calvin opts for the latter reading. He suggests that when the waters covered the whole earth during creation, they were more of a grave than a garment.

Calvin admits that the context leads us to a different view. The passage is describing creation, and thus the natural reading is that the verse refers to waters covering the earth during creation. He suggests that we should understand the psalm as being written in the "potential mood." That is, the text says not that the earth *was* covered with waters as a garment but that it *would be* covered with waters as a garment.

He suggests that this is vindicated by the grammar of the Hebrew language. Hebrew has a very different tense and mood system than the European languages we are familiar with. Translators have to infer the correct tense and mood based on the context and meaning of the verse. Because of that, Calvin can argue that it is possible that it means that the waters *would* cover the earth. However, in order for this to work, Calvin would have to show that the context supports that meaning.

Calvin interprets the statement that the waters stood above the mountains as indicating that they *would* stand above the mountains if God were not commanding the waters to stay away. The mountains are elevated and valleys sink below only because God has set bounds on the waters, preventing them from overflowing the earth.

Calvin takes the passages as indicating that the sea is a beautiful garment for the earth. However, the seas do not cover the earth. The inhabitable surface of the earth stands uncovered. But the psalmist describes it as a garment covering the earth as a metaphor because it only remains uncovered by the divine providence of God. If he were to cease his providence, the waters would rush down and cover the mountains.

Calvin suggests that the ascending of the mountains and the descending of the valleys are poetical figures. They suggest that unless God held back the seas, the distinction between mountain and valley would be erased because everything would be covered with water.

Conclusions

Historical Christians

Prior to about the year 1000 AD, almost all Christians thought that the seas were resting in basins. They suggested that either God created the basins during creation or somehow made the waters denser so they took up less room. They saw God's divine command as preventing the waters from rising up and overflowing the earth. The only exceptions are those, like Chrysostom, who thought that the earth was floating on the seas.

Towards the end of the Middle Ages, the idea took hold that the seas were above the land. Dante argued against that position. However, it was widely accepted by the time of the Reformation. Both Luther and Calvin interpreted the Biblical text as referring to these piled-up waters.

Modern Science

Christopher Columbus would discover the Americas in 1492. This showed that there was another landmass beside the one known to Europeans. Part of the impetus for the theory that the seas were above the earth was to explain why there was only one major landmass. Since there was another major landmass, one of the major reasons for the theory was undermined. As modern science replaced the classical conception, the idea that the waters ought to be a sphere surrounding the earth also fell by the wayside. Moreover, it also became more broadly understood that the illusions which make the sea look like it is above the land are caused either by the curvature of the earth or by the relative featurelessness of the sea.

Evaluation of the Response

The early Christians were correct to think that the oceans sat in basins. However, they did not take the position for Biblical reasons. Rather, it appears that historical Christians, the Biblical authors, and classical science all took this as a commonsensical assumption. Adopting alternative ideas, such as a floating earth or oceans at higher elevations, required more developed scientific theorizing.

Christians at the time of the Reformation were wrong to think that the oceans were above the land. While they did cite Biblical passages in support of this idea, it seems clear that they were actually simply following the science of the day and reading it into the text. If we look at how Calvin deals with passages that do not fit well with that theory, we find him forced to take some awkward readings of the text in order to accommodate the science of his day. Luther and Calvin both consider passages that contradict the theory and excuse these with claims that they are accommodated to our understanding.

This provides another example of the respect for Biblical authority amongst historical Christian authors. In the early authors, we see this in how they tried to determine what happened to the waters that had covered the earth during creation. They were taking the creation account seriously and sought to try to understand it. Even when the Reformers read flawed scientific ideas into the text, we see that they thought the text was speaking to scientific issues. They did not dismiss passages that seemed to contradict the theory but sought to explain them. Great fault can be found justly in their handling of the relevant texts, but they still thought that they needed to take those texts seriously.

The solution is, yet again, returning to the source. The flawed solution followed by Luther and Calvin was to read the science of their day into the text. This is not to say that they did so explicitly or deliberately. However, in practice, they read certain passages as teaching a flawed scientific view while dismissing other passages that contradicted that view on a flimsy basis. The solution, instead, was to go back to the Scriptures and see what they really sought to convey, as well as to go back to the scientific data to determine whether the scientific theory was true. By going back to the sources, it is clear that the scientific theory was false and the Scriptures do not teach it.

Chapter 12

Is There a Single Ocean?

The Scientific Question

The Middle East was the cradle of civilization. The world's earliest civilizations originated in this region. It would later be the economic and intellectual center of the Roman Empire. This region lies at the junction of Europe, Asia, and Africa. It also borders both the Mediterranean Sea and the Indian Ocean. But it was not clear whether the Mediterranean Sea and the Indian Ocean were distinct bodies of water or connected. Were they simply two parts of a vast, world-spanning ocean, or distinct bodies of water?

The idea that there was a single, world-spanning ocean has a long history stretching back into mythology. The oldest known visual depiction of the world comes from Babylon and depicts an ocean surrounding the earth. Greek myth speaks of Oceanus, a god associated with a river encircling the whole earth.

The conquests of Alexander the Great brought knowledge of the geography of the lands from the Middle East to India into the classical world. Furthermore, conquests and explorations into Europe brought knowledge of the Atlantic Ocean. The existence of vast oceans in either direction suggested that there was a single world ocean surrounding all of the land. Furthermore, there were reports, of varying plausibility, of ships sailing around Africa. Thus, it was taken as well established in the classical world that there was a single, vast interconnected ocean.

The Biblical Discussion

The Genesis 1 creation account describes the division of land and sea:

> And God said, "Let the waters under the heavens be gathered together into one place, and let the dry land appear." And it was so. God called the dry land Earth, and the waters that were gathered together he called Seas. And God saw that it was good. (Genesis 1:9–10 ESV)

The waters are said to be gathered into one place. This could refer to the ocean, but what about waters in lakes or rivers that are not part of the ocean? However, this gathering is also called the seas, using the plural. The Biblical text uses the word seas not only to refer to parts of the ocean, such as the Red Sea or the Mediterranean Sea, but also to certain lakes, such as the Salt Sea or the Sea of Galilee. Even mighty rivers such as the Nile or Euphrates are sometimes referred to as seas. How can the waters be said to be gathered together into one place and yet also into these many distinct bodies of water?

It may seem obvious to understand "one place" as a reference to the ocean. After all, the vast majority of all the water on earth is in the oceans. One could understand the reference to multiple seas as referring to different parts of the ocean. Thus, the waters could be described as gathered to a single place, the ocean, while still being found in multiple seas. However, this does not align with the Biblical usage of the term sea, which refers to any large body of water. Furthermore, there is no other Bible passage that suggests the existence of a single interconnected ocean that would be divided into multiple seas.

The text suggests that the author considered the diverse seas to together make up a single place. But in what sense are all these bodies of water a single place? One could maintain that almost all bodies of water flow into the sea, and are thus touching and can be considered one continuous place. Nevertheless, this would seem to require a rather unusual notion of a single place.

As discussed in a previous chapter, the Bible describes the waters as being under the earth. This suggests that the one place referred to here is likely simply "under the earth." It is a very general and vague notion of the place that includes all bodies of water. When the text refers to these waters' being gathered the one place, it means that while the waters and the earth were previously intermixed, now the waters are restricted to the area below the dry land.

Historical Christians

Basil of Caesarea

Basil considers the objection that the waters are not all gathered into one place.[1] He argues that the text is speaking of the chief or largest gathering of water. It is not concerned about puddles or other smaller bodies of water. Rather, it is speaking only of the most important body.

Basil draws an analogy between classical science's conception of layers of fire and air above the earth and the single location of the waters. While there is fire near the earth, the

---

[1] Basil of Caesarea, "Hexaemeron," Homily IV, Section 4.

main mass of the fire is far above in the realm of fire. Likewise, there is water in various places on the earth, but the main body is in the ocean.

Basil mentions many lakes and seas that contain water, but much less than the ocean. Furthermore, these bodies are connected to and drain into the ocean. He appeals to people who have travelled around the earth and to geographers who have shown that there is only one ocean. In particular, he argues that the Red Sea and the Atlantic Ocean only form one. Since they communicate or are connected, they all form part of one great gathering.

Why did the text say that God called the gathering of the waters seas? Basil explains that the different parts of the ocean are called different seas.

Ambrose of Milan

Ambrose asks how the waters could form one mass when they are observed to be spread among various bodies of water.[2] To answer this question, he says that we must weigh the words of Scripture with exactitude.

Ambrose argues that the various seas are all connected, forming one major ocean. He goes into detail, describing the connections between the various seas that would have been familiar to his audience. Everything is connected by the ocean, which surrounds all land. The various seas have different names but are all part of one watery mass. Ambrose draws an analogy to there being one earth and yet many lands. In the same way, there is one general sea and yet many seas.

In order to support his point, he references a couple of Bible passages that refer to the sea, suggesting that this is a reference to the ocean. One is Psalm 89:11–12 which does not refer to the

---

[2] Ambrose of Milan, "Hexameron," Chapter 3.

sea in the Hebrew text, but does in the Septuagint. Another is found in Job:

> Or who shut in the sea with doors when it burst out from the womb, (Job 38:8 ESV)

There are a few more Biblical passages that refer to the sea; however, these passages appear to generally be referring specifically to the major sea familiar to the Israelites, the Mediterranean Sea.

Ambrose acknowledges that there are waters that are not part of the main ocean. He mentions three lakes in Italy as examples. To explain this, he draws an analogy to God's creating the sun, moon, and stars. We say that God created the two great luminaries without implying that the stars do not exist. Likewise, just because the text only talks about the ocean does not mean that other bodies of water do not exist.

Bede

Bede explains that the waters are all joined by a continuous stream of water into the great ocean and the sea.[3] Even lakes that seem self-contained still flow into the sea, even if through hidden chasms. He thinks this is also demonstrated by wells, which he believes are supplied by the sea.

The waters are called seas because that is where most waters are. Bede also notes that the Hebrews call all gatherings of waters seas, whether they are saltwater oceans, salt lakes, or freshwater lakes. Thus, they are called seas in the plural because there are many bays or divisions of the general mass of waters.

---

[3] Bede, "Commentary on Genesis," Book One.

John of Damascus

John addresses this issue while discussing the gathering of the seas.[4] John points out that the text immediately states the waters were gathered into seas, and thus the waters were not gathered into one place. Instead, he argues that the waters were gathered into their various distinct collections. Thus were formed the Red Sea, the Mediterranean Sea, and the Indian Sea. Likewise, the lakes were also formed in this way.

Despite not attributing the gathering of the waters into one place as referring to a single connected ocean, John still believes that there is an ocean circling the earth. However, his information about the ocean is not correct. He thinks it is sweet and drinkable and that it provides water for the seas. Nevertheless, he gives a correct explanation for the reason that the seas are salty—the water has nowhere to go and the evaporation of waters leaves the salt behind.

Remigius of Auxerre

Remigius explains the reference to the one place where waters were gathered to be the single ocean.[5] Remigius suggests first that the word seas is used because the ocean is divided up into many seas, which have a variety of names. Alternatively, he suggests that the plural is used as Hebrew idiom, saying that Hebrews "call all gatherings of waters 'seas.'" But this is not correct because the Hebrew text does, in fact, use the singular for individual bodies of water.

Thomas Aquinas

Aquinas considers the objection that waters do not occupy the

---

[4] John of Damascus, "An Exposition of the Orthodox Faith," Book II, Chapter 9.
[5] Remigius of Auxerre, "Exposition on Genesis."

same place.[6] In particular, they are not in continuous contact and thus cannot be described as being in the same place. He suggests that this is because all waters have the sea as their goal. Alternatively, he suggests that one place might refer not to a single location as such but rather to a domain apart from the land. Further, this seems to be suggested by the idea that waters are distributed to multiple seas and not a single sea.

Martin Luther

Luther argues that Moses was speaking in a "plural or distributive sense."[7] That is, the text means that the waters were gathered to various places instead of one place. Thus, the waters were gathered onto various seas, rivers, lakes, etc. He does not elaborate on how this could be a justified reading of the text.

Conclusions

Historical Christians

Most authors took the one place referred to in Genesis 1 to be the ocean. This included Basil, Ambrose, Bede, and Remigius. They explain the reference to multiple seas as being a subdivision of the broader ocean into various seas that together make up the whole ocean. Other waters are included either because they flow into the sea or because they are smaller bodies of water that did not need to be mentioned.

John of Damascus and Luther both read the text as referring to a gathering into multiple bodies of water rather than a single

---

[6] Aquinas, *The Summa Theologiæ of St. Thomas Aquinas*, First Part, Question 69.

[7] Luther, *Luther On The Creation: A Critical and Devotional Commentry on Genesis*, 1:Chapter 1, Section II, Part III.

body of water. Neither goes into much detail as to why they thought it justified to read the text in that fashion.

Aquinas gives the common response, pointing out the interconnectedness of the ocean with all other waters. However, he also proposes that the one place is meant merely as being contrary to the land and not as constituting a single body of water.

Modern Science

The interconnectedness of all oceans was firmly settled by voyages around the world. Towards the end of the fifteenth century, Bartolomeu Dias sailed around Africa, reaching the Indian Ocean from the Atlantic. Ferdinand Magellan's expedition sailed around the world, traveling west through the Atlantic, Pacific, and Indian oceans at the beginning of the sixteenth century. While the existence of a single ocean had long been believed, these explorations definitively settled the question.

Modern science has also confirmed that almost all bodies of water flow into the ocean. However, there are exceptions. A few lakes, notably the Dead Sea, are below sea level and do not flow into the ocean. Water only leaves the Dead Sea through evaporation.

Evaluation of the Response

None of these authors challenged the idea of a single world ocean. They accepted the science of their day's conclusion that there was a single, world-spanning ocean. Rather, most of the authors took the text as affirming this idea. This scientific idea has turned out to be correct. The only definite scientific error is that of John of Damascus, who thought the ocean was filled with sweet and potable water. But this appears not to have come from the Bible but merely from incorrect knowledge about the world.

However, just because there is a single planet-spanning ocean does not mean that these authors were correct in taking the Scriptures to be referring to that ocean. Indeed, as argued earlier, it is most likely that the text is not talking specifically about the ocean when it refers to the "one place." Historical authors were somewhat aware of these issues as they sought to reconcile the reference to multiple seas with a single ocean. Actually, these authors were reading the science of their day into the text. This is a fairly innocuous case since the science turned out to be accurate and understandable given how readily "one place" seems like a reference to the ocean.

This provides another example of the way that historical Christians understood the authority of the Biblical text. They assumed that the Biblical text spoke accurately about the distribution of waters on the earth. None thought the Biblical account was inaccurate or could not be relied upon to answer this scientific question. Rather, because the text affirmed that the waters were gathered to one place, the waters were indeed gathered to one place, whatever that might mean.

We also see, again, the importance of returning to the source. It is important to go back and re-evaluate the text to determine what it actually says. There is a temptation to simply accept a flawed interpretation when it seems to fit the science of the day. But this is a mistake; we should hold our Biblical interpretation to a higher standard.

Chapter 13

Are Birds Made From Water?

The Biblical Discussion

The Genesis 1 creation account describes the creation of fish and birds:

> And God said, "Let the waters swarm with swarms of living creatures, and let birds fly above the earth across the expanse of the heavens." So God created the great sea creatures and every living creature that moves, with which the waters swarm, according to their kinds, and every winged bird according to its kind. And God saw that it was good. (Genesis 1:20–21 ESV)

However, the Septuagint describes the event somewhat differently:

> And God said, "Let the waters bring forth creeping things among living creatures and birds flying on the earth against the firmament of the sky." And it became so. And God made the great sea monsters and every creature among creeping animals, which the waters brought forth according to their kinds, and every winged bird according to kind. And God saw that they were good. (New English Translation of the Septuagint, Genesis 1:20–21)

The primary difference to notice is that while most modern English translations state that God commanded the waters to swarm with the living creatures, the Septuagint indicates that the waters are to bring forth the living creatures. This suggests that the birds and fish were made from the waters.

However, this is not the only difference. The Septuagint also references "creeping things" and adds the phrase "And it became so." All three changes appear to have come from the description of Day Six:

> And God said, "Let the earth bring forth living creatures according to their kinds—livestock and creeping things and beasts of the earth according to their kinds." And it was so. And God made the beasts of the earth according to their kinds and the livestock according to their kinds, and everything that creeps on the ground according to its kind. And God saw that it was good. (Genesis 1:24–25 ESV)

Somehow the phrases "bringing forth," "creeping things," and "and it was so" were copied from the description of Day Six into Day Four. It seems that the translator either deliberately borrowed from Day ix or that some of the copyists were not sufficiently careful about which text they were copying.

The Vulgate would follow the Septuagint:

> God also said: Let the waters bring forth the creeping creature having life, and the fowl that may fly over the earth under the firmament of heaven. And God created the great whales, and every living and moving creature, which the waters brought forth, according to their kinds, and every winged fowl according to its kind. And God saw that it was good. (Genesis 1:20–21 Douay-Reheims Bible)

A number of early English translations also follow the Septuagint and Vulgate, including the King James Version. However, most modern translations follow the Hebrew text instead and indicate that the waters swarm or are filled with living creatures, not that they were created from the waters.

The consequence is that throughout the historical period under consideration in this book, it seemed to almost all Christians that the Bible taught that birds were brought forth out of water.

This creates a conflict with a mention of the creation of birds later in the text:

> Now out of the ground the LORD God had formed every beast of the field and every bird of the heavens and brought them to the man to see what he would call them. And whatever the man called every living creature, that was its name. (Genesis 2:19 ESV)

This text indicates that God formed the birds out of the ground, just as he formed the land animals. This contradicts the claim that the birds were created out of water.

The Scientific Question

From the perspective of classical science, it made little sense that birds should be made out of water. Birds were associated with the air and not the water. It would make some sort of sense if fish were created out of water because they lived in it. But by the same token, it would only make sense if birds were created out of the air. Why would birds be created out of water like fish when they have very little in common with fish?

Alternatively, the element earth was thought to explain solidity. Everything that was solid or heavy was thought to contain earth. Since both birds and fish were clearly relatively solid and fell to the ground like earth, this suggested that they ought to have been made from earth and not water. If they had been made out of water, they ought to have been liquid or floating like water.

Historical Christians

Ephrem the Syrian

Ephrem understands the birds to have been created out of the water.[1] He describes birds soaring in flocks out of the waves into the air. He refers not to fish or sea creatures but to serpents, following the Septuagint's rendering: creeping things.

This is curious because Ephrem wrote in Syriac and thus would be expected to be less influenced by the Septuagint. Syriac is a dialect of Aramaic and thus similar to Hebrew, and Ephrem is typically less led astray by errors in the Septuagint than many others in the early Church. This can be seen in his quotation of Scripture, which is more akin to the Hebrew text than to the Septuagint. Nevertheless, it appears his reading was still influenced by the Septuagint.

Basil of Caesarea

Basil asks why the waters were commanded to bring forth these creatures?[2] What is the difference between them and the land creatures? He suggests that the creatures made from water are a lesser form of life. They have dull sight, no memory, imagination, or social interaction. He suggests that in aquatic animals, their bodies rule over their movements, but in land animals, the souls rule over the bodies.

Basil points out that the four-legged creatures have strong senses and a memory of past events. Basil suggests that God created only the body for the sea creatures, but created both a

---

[1] Ephrem the Syrian, *Commentary on Genesis*, Chapter 1.
[2] Basil of Caesarea, "Hexaemeron," Homily 8, Section 1.

body and a ruling soul for the land animals. He acknowledges that land animals do not have reason. But they express their emotions. Sea creatures are, he insists, dumb. They cannot be tamed, taught, or trained. An ox knows his master, but a fish knows nobody. He discusses camels' showing resentment and argues that no sea creature can do the same.

Basil explains birds and fish's being created out of the waters because they have a "family link." He points out a similarity in their movements: fish swim through waters in much the same way that birds fly through the air. They are able to move vertically through the water or the air. In contrast, most land-dwelling creatures are stuck on the surface of the ground.

Ambrose of Milan

Ambrose draws parallels between the swimming of fish and the flying of birds.[3] A bird cuts through the air just as a fish cuts through the water. A fish controls his path through his fins, just as a bird controls his path through his wings. Ambrose states that this similarity is due to both having water as their origin.

Augustine of Hippo

Augustine considers the objection that it makes no sense that the birds would be made from the waters.[4] However, Augustine indicates that he considers this a misrepresentation of what the text says. Augustine argues that the waters can refer to moist air. This air is filled with vapors coming from the sea and the land. This is necessary in order to support the flight of the birds. He further states that these vapors are responsible for dew.

---

[3] Ambrose of Milan, "Hexameron," Homily 8, Chapter 14.

[4] Augustine, "Two Books on Genesis against the Manichees," Book 1, Chapter 15.

Augustine appeals to "learned men" for the idea that the term waters can refer to moist air. He does not refer to any other Biblical passage that uses water this way. He cannot, for no such passage exists.

Augustine appeals to experience with Olympus, a tall mountain in Greece, where it is claimed that there was no wind, cloud, rain, or birds. According to Augustine, it was so high that it was above the region of moist air. This was attested to by the reports of people who climbed the mountain to offer sacrifices. This is incorrect, as there is, in fact, rain on Olympus and birds that fly higher than any mountain. Augustine argues that it makes sense for birds to have been created from water because they only fly through the moist portion of the air.

Augustine returns to this subject in his later commentary (*The Literal Meaning of Genesis*, 3:6–7). He gives a similar answer, but adds that this is evidence of divine knowledge on the part of the author.

Bede

Bede points out the conflict between understanding Genesis 1 to say that the birds were made from the waters and understanding Genesis 2 as indicating that they were made from earth.[5] He proposes two possible explanations for this. First, he suggests that Genesis 2 does not actually state that birds were made from the earth but only makes that statement about land animals. This means reading the text as though it said that God formed the land animals out of the earth and also formed the birds, but it does not specify the material from which birds were created. This reading is difficult to justify grammatically. Bede's second suggestion is that the earth is used in a broader sense, corresponding to the whole terrestrial realm, as in the

---

[5] Bede, "Commentary on Genesis," Commentary on Genesis 2:19.

phrase "heavens and earth." Thus, the birds are made of earth in a broader sense, even if they were created from water.

He supports this reading by reference to the Psalms:

> Praise the LORD from the earth, you great sea creatures and all deeps, (Psalm 148:7 ESV)

He observes that the sea creatures are mentioned as though they are part of the earth. Thus, Bede proposes that the word earth is meant to be understood in a broader sense than just the dry land.

However, in the Hebrew text of Genesis 2, the birds are formed from the ground (אֲדָמָה *adamah*), not the earth (אֶרֶץ *erets*). These two words have similar meanings, but while *erets* is used to refer to earth and sky together, *adamah* isn't. *Adamah* more specifically refers to the ground. Bede was probably unaware of this because both the Septuagint and the Vulgate translated the word as simply earth.

Thomas Aquinas

Aquinas considers the objection that neither birds nor fish were made from water.[6] Clearly, their bodies naturally fall to and rest upon the earth. They do not, like water, float on top of the earth. Aquinas grants that they are made mostly of the element, earth. However, he argues that they are described as being created from water because they have a special affinity for the bodies in which they move. This explains why fish would be made out of water, but probably not birds.

Aquinas considers the objection that birds should have been made from the air and not the waters. He explains that the author did not mention the air since it was not apparent to the senses. Indeed, Hebrew lacks a word that corresponds

---

[6] Aquinas, *The Summa Theologiæ of St. Thomas Aquinas*, The First Part, Question 71.

to our English word, air. Aquinas claims that the text some-times refers to the air as water and sometimes as the heavens. Because of this, it is appropriate to describe the waters as the origin of birds.

Martin Luther

Luther reads the text as declaring that the birds and fish have a common origin in the waters.[7] He links this to a similarity in movement: fish swim through the sea like birds fly through the air. Even though their flesh differs, they have the same nature.

Luther approves the opinion of physicians who prefer the flesh of birds to that of fish. Birds live in a purer element, the air, and are thus to be preferred. However, Luther contends against the philosophers who find the common origin of birds and fish objectionable. He insists that we must take the Scripture against the opinion of philosophers.

It goes against our reason to think that birds might be made from the waters. How can a creature be created out of water that cannot live in it? Nevertheless, the Word of God shows that for God, all things are easy. The creation of the birds from water shows the divine authority and majesty of the Bible.

John Calvin

Calvin brings up this objection, and he accepts that the Bible does indicate that birds owe their origins to the waters.[8] He simply says that God could make birds out of water if he wanted to. If he can create everything out of nothing and light out of darkness, he will have no difficulty creating birds out of water. He points out that if this is insufficient, we can point out that water has a greater affinity for the air than the earth does.

---

[7] Luther, *Luther On The Creation: A Critical and Devotional Commentry on Genesis*, 1:Part V.

[8] Calvin, *Calvin's Commentary on the Bible*.

Conclusions

Historical Christians

Ephrem, Basil, Ambrose, Bede, Luther, and Calvin all conclude that the text teaches that birds were created from the waters. Most commonly, this was explained by pointing to the similarity between flying through the air and swimming through the ocean. Only Bede took notice of the conflict between this understanding and the Genesis 2 account.

Augustine and Aquinas argue that there is moisture in the air, suggesting either that the birds were made from this wet air or that it makes sense to attribute the origin of birds to water because there is also water in the air. Aquinas argues that this is because we cannot see the air, and thus the author of Genesis refers to it as the waters. But we have no indication that the ancient Hebrews conceived of the lower atmosphere as wet air or part of the waters. This is not very plausible exegesis.

Modern Science

Modern science abandoned the classical system of elements, rendering the scientific question largely moot. Earth, air, and water are no longer considered elements. Nobody would argue that birds must be made of earth because they are heavy and solid. Nor would anyone argue that the birds should be made from the air in which they fly.

With modern science, we understand that living creatures are made up of a variety of elements that are commonly found in our environment. Consequently, it would be possible to find all of the constituent elements necessary to create a living being in either the earth's crust or the sea. As such, there is no clear reason within modern science to favor earth, atmosphere, or the seas as the origin of elements that God would have used to create the birds.

Evaluation of the Response

Most historical Christians believed that the Scriptures taught that birds were created from water. But this was based on a flawed translation of the Hebrew text. Thus, the point is now largely moot. At the same time, the scientific questions raised in the minds of historical Christians by this are also moot since they were rooted in a classical conception of the elements. The scientific and Scriptural issues are thus moot.

The mootness of the issue actually provides reason for encouragement. What was once an objection against the veracity of the Biblical text is now moot. The objection fell by the wayside with better knowledge about the Scriptures and science. This demonstrates that, while they are easily forgotten about, there are cases where apparent scientific difficulties have been resolved.

This issue also demonstrates the historical Christian respect for the authority of the Biblical text. We now understand that the Bible does not indicate that birds were made from water. However, historical Christians *thought* that the Bible stated that the birds were made from water. They insisted on holding to what they thought the Bible taught on this scientific question. Some, such as Aquinas and Augustine, did attempt to argue that the text was referring to moist air rather than water, but they were arguing that this is what the Biblical text intended to say.

The solution was to go back to the source. By returning to the original Hebrew text, it is clear that the Bible does not claim that birds were made out of water. By re-evaluating the scientific data, we realize that the classical system of elements was in error. We see that adopting doubtful readings, such as

238

suggesting that the Bible was referring to moist air, was unnecessary and unhelpful.

Chapter 14

Was There a Single Spring Watering the Whole Earth?

The Biblical Discussion

At the beginning of the second account of the creation of man, Genesis says:

> When no bush of the field was yet in the land and no small plant of the field had yet sprung up—for the LORD God had not caused it to rain on the land, and there was no man to work the ground, and a mist was going up from the land and was watering the whole face of the ground— (Genesis 2:6 ESV)

What is this "mist" that goes up and waters the whole face of the ground? The Septuagint says something a little different:

> yet a spring would rise from the earth and water the whole face of the earth. (Genesis 2:6 New English Translation of the Septuagint)

The Septuagint says that a spring rose out of the ground instead of mist. The Hebrew word אֵד (ed), translated as mist or spring, has an unclear meaning. It shows up in only one other biblical passage, where it is referring to the source of rain:

> For he draws up the drops of water; they distill his mist in
> rain, (Job 36:27 ESV)

Furthermore, the Targums, an Aramaic translation and commentary of the Hebrew Bible, translate it as cloud. This suggests that what is being referred to is simply clouds, the source of rain. On the other hand, the Septuagint translated the word as spring. However, a similar word is found in the Akkadian language (which is related to Hebrew) that is argued to refer to subterranean waters, lending support for the translation: spring.

The text begins by stating that the various plants of the field have not sprung up because God has not sent rain and there is no man to work the ground. Next, the "mist" waters the ground, and then God forms man. The "mist" is the solution to the lack of rain, just as creating Adam is the solution to the lack of a gardener. The most plausible solution to a lack of rain is rain, whereas it is possible but less plausible that another irrigation method was used instead as a substitute.

The text says that the mist or spring was going up from the ground. If mist or rain is intended, this could be a reference to evaporation. One might be inclined to be skeptical of the author's being scientifically literate enough to refer to evaporation. However, we have a number of Biblical passages that appear to refer to evaporation:

> And at the seventh time he said, "Behold, a little cloud like
> a man's hand is rising from the sea." And he said, "Go up,
> say to Ahab, 'Prepare your chariot and go down, lest the rain
> stop you.' " (1 Kings 18:44 ESV)

> He it is who makes the clouds rise at the end of the earth,
> who makes lightnings for the rain and brings forth the wind
> from his storehouses. (Psalm 135:7 ESV)

> When he utters his voice, there is a tumult of waters in the heavens, and he makes the mist rise from the ends of the earth. He makes lightning for the rain, and he brings forth the wind from his storehouses. (Jeremiah 10:13 ESV)

Alternatively, it might be that there are underground waters rising up through the spring. However, we do not have other passages describing springs in that way. Rivers and wells are sometimes said to rise up, but not springs.

The text says that the whole face of the ground was watered. This makes sense if the text is referring to rains or mists, which could easily make the whole ground wet. However, it makes less sense if the text is referring to a spring or river, which would normally only water the ground next to the spring or river.

Altogether, these lines of evidence favor the idea that the text is referring to ordinary rain. Nevertheless, there are some arguments to be made for the idea that the text is referring to a spring. Whatever the intention of the Hebrew text, the Septuagint, as mentioned, indicates that it was a spring. The Vulgate followed this. However, with the Reformation came new translations directly from the original text. These translations translated this verse using mist or other similar words. More recent scholarship has questioned this, and some of the most recent translations translate it as springs, stream, streams, or just simply water.

Regardless of the nature of the waters, they are said to water the whole face of the ground. Both the Septuagint and the Vulgate state that it covers the whole face of the earth. Did the author intend to convey that the whole planet was watered by this single spring or mist?

The passage opens by stating that it had not rained on the (אֶרֶץ *erets*). The word is often translated as earth, although in this passage, the ESV translates it as land. This is because the word sometimes refers to the entire planet, but in other cases, it refers to a geographically restricted area. A possible interpretation is that the text is speaking of the area around Eden, and thus it is the whole face of the ground of that region that is watered and not the face of the whole earth. This makes sense in light of the fact that the Genesis 2 creation account focuses on the Garden of Eden.

Today, there is some debate about the intended meaning of this text. It may simply refer to rain, and that in a specific geographic area. Alternatively, it may refer to springs or streams coming up from underground and watering a wide area. Whatever the intended meaning, in the Septuagint and Vulgate translations, the text claims that a single spring watered the whole earth.

The Scientific Question

There is no spring that waters the entire earth. Furthermore, such a spring would require an enormous amount of water in order to be able to water the whole earth. It is very difficult to see how the waters could reach high elevations like mountains. Such a spring thus does not and could not exist.

In the ancient world, it was believed that most of the water in rivers came from underground springs fed by the ocean. Water from the ocean was thought to work its way up through the ground into various springs all around the earth. A large part of the reason for this was that it seemed as if there was insufficient rain to account for the large volume of water observed constantly pouring into the sea. As such, while there was not a single spring watering the whole earth, they did think that there were many springs watering the ground.

Historical Christians

Ephrem the Syrian

Ephrem understands these waters as coming up through the fountain from the great primordial deep and irrigating the whole earth.[1] There is no indication that he sees any potential issue with this.

Augustine of Hippo

Augustine asks what sort of spring was able to irrigate the whole face of the earth.[2] Augustine elaborates on the necessity of showing that there is no absurdity or contradiction in the text. The goal is to prevent people from giving up the faith or not approaching the faith because they think the events described in the Bible are impossible. We need to show how it is possible that the Biblical statement could be true. He calls on those who find the claim impossible to show another way in which the truth of Scripture can be understood. He insists that even if we cannot show Scripture to be true, it is still true.

His first suggestion is that the spring was removed as punishment for man's sin. Once this spring was removed, growing plants became much harder, therefore multiplying man's toil. However, Augustine rejects this possibility because, a few verses later, the text describes four rivers that water the earth. It does not make sense for the earth to be watered both by this single massive spring and by these four rivers.

Augustine then suggests that the spring was a temporary feature during God's process of creation. God originally watered the entire earth using this spring, but afterwards either

---

[1] Ephrem the Syrian, *Commentary on Genesis*, Chapter 2.
[2] Augustine, *The Literal Meaning of Genesis. 1*, Book 5, Chapter 7, Section 21-24.

removed the spring or greatly reduced it to become the source of the four rivers.

However, even if the spring no longer exists, it is still difficult to see how it might have watered even the tops of mountains. Augustine appeals to places with recurring floods, most famously the Nile in Egypt, to argue that a single spring might have watered the whole earth in a similar way. The spring could have flooded the whole earth before allowing the waters to recede. Alternatively, he suggests that the brand new earth was much more level and thus much more easily watered.

Augustine then suggests another explanation: that the text is referring to numerous springs all around the earth instead of a single spring. He describes the waters as coming out of the earth like locks of hair coming out of a head. He suggests that all of these springs could be described as a single spring because they were united in some sense. In particular, he suggests that they were all connected through water in underground caverns. He goes so far as to claim that only a contentious person would reject this reading.

Ultimately, Augustine deems the most likely explanation to be a grammatical one. As he describes it, the text is using the singular for the plural. When the text refers to "a spring," it really means "springs." Augustine appeals to a couple of passages from the Psalms in support of reading the singular as the plural:

> He sent among them swarms of flies, which devoured them, and frogs, which destroyed them. (Psalm 78:45 ESV)

> He spoke, and the locusts came, young locusts without number, (Psalm 105:34 ESV)

In the Septuagint and the Latin translations that Augustine was using, the references to frogs and locusts are in the singular, following a literal translation of the Hebrew. Today, most translators take the Hebrew words for frog or locust to be plural by definition, akin to English words like rice, sheep, and clothes, and thus translate them in the plural.

But there is a basis for Augustine's reading in an idiom called the definite generic. Examples of this idiom include:

- The dog is man's best friend.
- The statesman stands for what is right rather than what is popular.
- The television revolutionized entertainment.

In each case, we use the definite article, the, and the singular noun to refer to the archetype or ideal of something. In the case of man's best friend, the reference is not to a specific dog but to the archetypical dog or the generic idea of dogness. The Biblical text uses this idiom in some places:

> And I will send hornets (the hornet) before you, which shall drive out the Hivites, the Canaanites, and the Hittites from before you. I will not drive them out from before you in one year, lest the land become desolate and the wild beasts (the wild beast) multiply against you. (Exodus 23:28–29 ESV)

The ESV translates the text as referring to hornets and wild beasts, but the Hebrew literally refers to "the hornet" and "the wild beast." However, the text does not refer to "the spring" but simply to a "spring," making it difficult to argue that it is an instance of the definite generic.

Augustine attempts several explanations for the single spring. He suggests that God may have originally created a single spring to water the whole earth, but that the spring no longer exists. Even so, it is difficult to account for how this spring would have watered the whole earth. Alternatively, he attempts to argue that the text could be speaking of many springs, even if it only literally refers to a single spring. But the arguments for this are not very compelling.

Bede

Bede's discussion of the issue borrows extensively from Augustine's, even including direct quotations.[3] To further support the idea of the spring temporarily flooding the earth, he adds a reference to the Jordan River, whose flooding is mentioned in Joshua:

> and as soon as those bearing the ark had come as far as the Jordan, and the feet of the priests bearing the ark were dipped in the brink of the water (now the Jordan overflows all its banks throughout the time of harvest), (Joshua 3:15 ESV)

Bede points to another passage that describes the nearby plain as well-watered:

> And Lot lifted up his eyes and saw that the Jordan Valley was well watered everywhere like the garden of the LORD, like the land of Egypt, in the direction of Zoar. (This was before the LORD destroyed Sodom and Gomorrah.) (Genesis 13:10 ESV)

Bede suggests that the Jordan sometimes floods to such an extent that it waters the nearby plain, which is why the cities were once in a highly fertile state instead of a desert.

Bede repeats Augustine's argument that the singular might mean the plural. However, he introduces an error in that he describes Augustine's example of "soldier" to mean "soldiers" as Scriptural. Augustine used the word soldier as an example of the principle of the definite generic in common usage, not as

---

[3] Bede, "Commentary on Genesis."

a Scriptural example. However, Bede appears to have misunderstood him.

Remigius of Auxerre

Remigius closely follows the discussion of both Augustine and Bede.[4] He differs somewhat in his choice of Biblical examples to support the idea of using the singular for the plural. He references the same version from the Psalms:

> He sent among them swarms of flies, which devoured them, and frogs, which destroyed them. (Psalm 78:45 ESV)

However, instead of focusing on the frogs, Remigius focuses on the flies. The same situation applies in that the Hebrew word is plural by definition.

He adds a suggestion that prior to the flood, there were not so many large mountains because he thinks that mountains were heaped from soft, loose earth during the flood. This would help explain how a single spring could have flooded all the earth.

Martin Luther

Luther wrote in his commentary on Genesis:[5]

> There was not as yet any rain, Moses says, to water the earth; but a certain mist went up and watered the whole face of the earth, to cause it to bring forth more abundantly afterwards.

Luther, without much commentary, describes it as a mist watering the whole face of the earth.

---

[4] Remigius of Auxerre, "Exposition on Genesis," Chapter Two.
[5] Luther, *Luther On The Creation: A Critical and Devotional Commentry on Genesis*.

John Calvin

Calvin understands the passage as describing a vapor that waters the earth which is distinct from either the rain or human irrigation.[6]

Conclusions

Historical Christians

Ephrem does not see a problem with postulating a spring that covers the whole earth. Augustine, Bede, and Remigius all see a problem and give a very similar discussion. They suggest that this spring no longer exists, that it might have temporarily or periodically flooded the whole earth, that numerous springs were united through underground waters, or that the Scripture was using the singular for the plural. The later authors, Calvin and Luther, used translations from the original texts, which indicate that a mist or vapor watered the ground and thus do not have to explain this spring.

Modern Science

The first thinker known to have thought that rain was the source of all rivers was Bernard Palissy, who lived at the time of the Reformation. In the sixteenth century, Pierre Perrault wrote a book called "On the Origin of Springs." He was able to show that there was sufficient rain to explain the source of water in rivers.

Springs are not fed from the ocean. Instead, the water in springs comes from water traveling from higher elevations through the earth. A spring is effectively sourced from an underground river or stream. Water does not generally rise up

---

[6] Calvin, *Calvin's Commentary on the Bible.*

out of a spring but flows in accordance with gravity. However, there are exceptions when the source of water is under pressure.

It remains the case that there is no single spring responsible for watering the whole earth. Moreover, there is also not a collection of springs watering the whole earth. The earth is watered by rain, and both springs and rivers consist of this water returning to the ocean.

Evaluation of the Response

The idea that a single spring watered the entire earth seems as implausible today as when the historical authors considered it. The earlier historical authors saw the problem but did not have a good solution. They suggested a number of solutions, but none of them was very compelling. Indeed, the fact that historical authors repeated the ensemble of solutions suggests that they did not really think that any individual solution was that compelling.

Nevertheless, we see that these authors respected the authority of the text. Even though they were certain that there was no such single world-watering spring, they thought they had to follow the text. They did not attempt to allegorize the text to avoid this idea. Rather, they attempted to come up with an explanation of how the text's description of a single earth-watering spring could work. Failing that, they tried to explain how the text could mean something different.

But we also see the influence of the science of the day on their interpretation. In their day, they thought that water made its way from the ocean up through springs all over the earth. This inclined them to read the text as referring to that. It prompts speculation that this idea may have influenced the Septuagint's translation of the Hebrew text. Whether or not it influenced the translation, it did influence how historical Christians interpreted that translation.

We can also take encouragement from this issue. To historical Christians, it seemed that the Bible made a seemingly incorrect scientific statement for which nobody had a good explanation. However, it is now clear that the Bible did not actually make that statement. It was simply an error in translation.

As we have seen, the solution was going back to the source. By going back to the original Hebrew instead of either the Septuagint or the Vulgate, we see that the text does not claim that a single spring watered the whole earth. Attempts to resolve the issue by appealing to doubtful exegesis, such as trying to interpret singular words as if they were plural, turned out to be unhelpful and unnecessary.

Chapter 15

Did Four Rivers Flow From Eden?

The Scriptural Question

The Genesis 2 account describes four rivers flowing out of Eden into surrounding lands.

> A river flowed out of Eden to water the garden, and there it divided and became four rivers. The name of the first is the Pishon. It is the one that flowed around the whole land of Havilah, where there is gold. And the gold of that land is good; bdellium and onyx stone are there. The name of the second river is the Gihon. It is the one that flowed around the whole land of Cush. And the name of the third river is the Tigris, which flows east of Assyria. And the fourth river is the Euphrates. (Genesis 2:10–14 ESV)

Upon first reading, this seems to be a straightforward description of the real-world geography of the ancient Near East. The Tigris and the Euphrates are the two major rivers that define the boundaries of Mesopotamia. However, problems quickly arise in trying to align the rest of the description with real-world geography. Cush is the Biblical name for Ethiopia, but the river said to encircle Cush, Gihon, is a spring in Jerusalem. The identity of the river Pishon is entirely mysterious. It runs through the land of Havilah, later identified as being the home of Ishmaelites and Amelakites, probably in the Arabian Peninsula, which lacks any major rivers.

A common approach historically was to identify the Pishon and Gion with significant rivers of the ancient world. Josephus, an influential Jewish writer of the first century AD, identified the Pishon with the Ganges, a major river in India, and the Gihon with the Nile.

The Bible mentions the Nile in many places, usually calling it יְאֹר (yeor), but never Gishon. This passage does identify Gishon with Cush, or Ethiopia, one of the sources of the Nile. However, it would be unexpected for the Biblical author to identify the Nile with Cush rather than Egypt.

The basis for the identification of Pishon with India is that the land of Havilah refers to India. This is largely on the basis of the identification of that land as having gold, bdellium, and onyx. In the ancient period, India was seen as an exotic and rich country, thereby seeming to fit the description of Havilah. Some authors particularly cite Pliny the Elder as indicating that India was full of gold. However, Pliny's account is not that India was naturally filled with gold but that much gold was sent to India in return for Indian goods.

Many modern translations refer to the land of Havilah having bdellium, which is a resin extracted from a tree that grows commonly in northern India. However, it is unclear that this is what the Hebrew refers to. The Hebrew word is (בְּדֹלַח) bedolach, and only appears in this verse and in Numbers 11:7 where its appearance is compared to manna. The Septuagint translates the word as carbuncle, a red gemstone. Others have suggested it might be a reference to pearls.

There is little reason to accept these identifications. Moreover, the Nile and Ganges do not flow from a common source, either with each other or with the Tigris and Euphrates. Even the Tigris and Euphrates do not have a common source. Quite the opposite; in ancient times, they both flowed into the Persian Gulf. Today, they merge into a single river, which then flows into the Gulf.

Changes in geography over time might explain some of the discrepancies. Rivers do change course or dry up. The boundaries of lands shift over time. However, the necessary changes seem too extreme for this explanation to work. The river of Gihon cannot plausibly have become a spring in Jerusalem. The Tigris and Euphrates derive from different mountain ranges and cannot plausibly ever have had a single source. The only way to make the geography described in this passage line up with known geography would be to appeal to such drastic modification that these are not the same rivers and lands we know today in any meaningful sense.

Another possibility is to appeal to ambiguity in the names of these rivers and places. The same name showing up in diverse geographic locations is common. For example, there are 186 places named Riverside in the United States. This is not anything particularly unusual; it is simply that Riverside is such a generic name that many different people have chosen it. Likewise, some of the names appearing in this passage are fairly generic. Within the table of nations, the name Havilah appears twice, with two different fathers:

> The sons of Cush: Seba, Havilah, Sabtah, Raamah, and Sabteca. The sons of Raamah: Sheba and Dedan. (Genesis 10:7 ESV)

> Ophir, Havilah, and Jobab; all these were the sons of Joktan. (Genesis 10:29 ESV)

A Benjamite named Cush is mentioned in the Psalms:

A Shiggaion of David, which he sang to the LORD concerning the words of Cush, a Benjaminite. O LORD my God, in you do I take refuge; save me from all my pursuers and deliver me, (Psalm 7:1 ESV)

Some of the names in this passage are generic enough that they could plausibly be used in multiple places. Others appear to be foreign words borrowed into Hebrew and difficult to explain in this way. It also leaves the question of why the author did not make it clear that he was not referring to the same lands and rivers his audience would have known but to other lands that just happened to have the same names.

The chief difficulty is that, however one identifies the four rivers, they all have a single source in Eden. Some have attempted to avoid this by arguing that the four rivers have Eden as a destination, not a source. The Tigris, the Euphrates, and various other rivers all join together as they head into the Persian Gulf. If the text were referring to these four rivers converging rather than diverging, it would be much easier to reconcile with known geography.

However, the text gives three indications that Eden is the source. Firstly, the rivers are said to "go forth" and not come into Eden. Secondly, the waters are described as being divided, not joined. Thirdly, the divided river is described literally as becoming four heads, typically taken to refer to the sources of the four rivers. It is difficult to defend the idea that describing a river that goes forth, is divided, and becomes four heads is actually describing rivers coming together.

I propose an alternative interpretation: this one river was taken from its place, scattered over the ancient world, and became, in some sense, four major rivers of the ancient world. This would actually be a more strictly literal reading of the text than understanding it as referring to one river branching into four. The first word of the key phrase is translated, "and from there," and almost always refers to the location of the subject before, with the implication that it is no longer in the same place afterwards. The second word is translated as "it was divided" and usually refers to something that is cut, scattered, or physically separated. The third word is translated as "and became," which usually refers to something that becomes true at a point in time. As such, the most literal reading of this phrase would indicate that the first river was scattered and replaced by four new rivers.

The text literally states that the river became four heads. The Hebrew text uses the term head in a variety of ways, including a literal head, the top of something, and a leader. But it does not ever otherwise refer to the source of something. As such, the most likely meaning here is that it refers to four chief or major rivers.

If this interpretation is correct, the author is drawing a contrast between the geography of the original creation, with one major river coming out of Eden, and the geography of his own day, which contained various lands and rivers. He does not claim that these four major rivers share a common source, and thus this interpretation avoids the difficult geographic problems. It also fits with the themes we see in the early chapters of Genesis, as a single family speaking a unified language becomes diverse nations speaking a myriad of tongues.

However, I must acknowledge that I am not entirely qualified to be proposing radical new interpretations of the Biblical text. This is, I think it is an intriguing possibility that is worth further consideration. But our purpose here is primarily to understand the way people have understood this text historically. Those who understand this as a historical account will typically invoke some combination of these different approaches. Some will offer alternative interpretations of the text. Some will appeal to a geography modified either by time or the flood. Some will argue that the names in the text are generic names reused for different places.

Historical Christians

Ephrem the Syrian

In his commentary on Genesis, Ephrem discusses the rivers.[1] Ephrem identifies the four rivers with rivers known today. He identifies the Pishon as the Danube, a river flowing through Europe that was, for a long time, the frontier of the Roman Empire. The Gihon he identifies as the Nile, the famous river of Egypt.

He argues that while the sources that we can see do not flow from Eden, they do flow from the Garden of Eden. He thinks the garden is located elsewhere and at a great height. The water from that garden is thus able to flow down under the sea and out from the location of the heads of these rivers.

Ambrose of Milan

Ambrose identifies the Pishon as referring to the Ganges, the major river of India. He identifies the Gihon as the Nile.[2]

Augustine of Hippo

Augustine insists that we should take this account in the literal sense.[3] That is, these are real rivers that flow from paradise. In particular, he points out that these are well-known rivers, showing that the whole creation is intended to be taken as a

---

[1] Ephrem the Syrian, *Commentary on Genesis*, Chapter 2.
[2] Ambrose of Milan, "Hexameron," Chapter 3.
[3] Augustine, *The Literal Meaning of Genesis. 1*, Book 8, Chapter 7, Section 13.

narrative in the literal sense. However, he notes that they may also have a figurative meaning as well.

Augustine explains that some of these rivers have changed their names. The Gihon is now known as the Nile, while the Pishon is now known as the Ganges. He supports this by referencing a change in the name of the Tigris from an older name, the Albula. Here, Augustine follows other ancient writers who recounted this change in name, although modern scholars aren't sure whether such a change actually took place.

Augustine explains that we may hesitate to take the account literally because of the source of the rivers. He argues that we should believe that the river does divide in paradise and flow under the earth to spring forth as the sources of these four rivers. He notes that rivers are known to run underground, although only for short courses.

Bede

Bede identifies the Pishon as the Ganges and the Gihon as the Nile.[4] He explains that these four rivers have known sources quite distinct from each other. Yet, the scriptures indicate that their source is in paradise. Bede argues that the waters must go under the earth and then burst forth in the various distant regions we know of as their sources.

Bede appeals to ancient geographers to show that rivers frequently flow into the earth and re-emerge in another location. The Lord did this to show us the way the rivers flow back to their true source in paradise. Bede is correct; there are underground rivers, but they are relatively rare. It certainly isn't the case that major rivers frequently disappear and re-emerge, as he suggests.

---

[4] Bede, "Commentary on Genesis," Book 1.

Bede identifies Havilah with a region of India populated by the descendants of Havilah.

> Pliny the Elder relates that regions of India are richer than other lands in veins of gold. Their islands are called Chryse and Argyre from their supply of gold or silver.

Bede makes reference to the same idea that Pliny indicates that there is much gold from India as well as the two legendary islands. He also supports the identification of Havilah with India by the presence of bdellium. He further points to the presence of onyx. He notes that other locations also have onyx, but argues that the Indian version has a sparkle.

Bede notes that no account is given of the Euphrates. This is because that river is near the promised land. As such, there was no need to explain it.

John of Damascus

John of Damascus identifies the Pishon with the Ganges and the Gishon with the Nile.[5]

Remigius of Auxerre

Remigius identifies the Pishon as the Ganges.[6] Remigius refers to the table of nations to identify Havilah:

> To Eber were born two sons: the name of the one was Peleg, for in his days the earth was divided, and his brother's name was Joktan. Joktan fathered Almodad, Sheleph, Hazarmaveth, Jerah, Hadoram, Uzal, Diklah, Obal, Abimael, Sheba, Ophir, Havilah, and Jobab; all these were the sons of Joktan. (Genesis 10:25–29 ESV)

---

[5] John of Damascus, "An Exposition of the Orthodox Faith," Book II, Chapter 9.
[6] Remigius of Auxerre, "Exposition on Genesis."

According to Remigius, the descendants of Havilah occupied a region in India that was still called by his name. He argues that Havilah refers to India on the basis of India having more gold than other lands. He cites Pliny the Elder for this.

Remigius identifies the Gihon as the Nile. He notes that the audience did not need to hear about the location of the fourth river, the Euphrates, as they already knew about it.

Remigius asks how these sources could flow from the Garden of Eden since we know they come from sources in our realm and not from paradise. He notes that the four rivers have four known and distinct sources. He gives the answer that they flow under the earth and burst forth at their appointed places. He notes that some rivers are known to do this. He briefly notes that some authors think that paradise is not a physical location, but he insists that because it has rivers and trees, it must be a physical place.

Thomas Aquinas

Aquinas brings up the objection that the rivers have sources that are clearly not in paradise.[7] The objection concludes that we ought not to take the description of the Garden of Eden as a physical place. Aquinas responds by quoting Augustine, arguing that the waters from paradise flow underground to become the visible sources of these rivers.

Martin Luther

Luther identifies this as one of the most difficult passages in the writings of Moses.[8] Luther accepts the identification of the Pishon with the Ganges and the Gihon with the Nile. However, he notes that they do not have a common source. The Tigris

---

[7] Aquinas, *The Summa Theologiæ of St. Thomas Aquinas*, First Part, Question 10.

[8] Luther, *Luther On The Creation: A Critical and Devotional Commentry on Genesis*, 1:Vol I, Chapter II, Part II.

and the Euphrates come from the north, while the Nile comes from the south. How can these statements then be reconciled with the facts?

Luther considers another possible explanation: that Eden was the entire world. If Eden contained the whole world, then the source of all rivers, regardless of where they are in the world, could be said to be in Eden. However, Luther insists that such a conjecture both is false and does not resolve the issue. These rivers do not merely have their source in Eden; they all have one source. Luther argues that God might well have enlarged Eden eventually if Adam had not sinned, but this does not give us reason to conclude that Eden must have covered the whole earth initially. Further, the Bible clearly distinguishes between Eden and the rest of the earth. He disparages Origen and those who would engage in allegorical understanding. He also heaps scorn on the commentator who pretends that there is no difficulty in this passage.

Luther argues that the garden was closed after the fall of man and then swept away by the flood. The flood destroyed all things. Not one trace of the prior world remains. He further argues that the flood would likewise have broken up and confounded the fountains of the rivers. The flood was a mighty convulsion that rearranged everything. He proposes that places such as the Mediterranean Sea, the Persian Gulf, the Arabian Gulf, and the Red Sea were not seas as they are now. Luther argues that the rivers are not in their former positions or flowing from their former sources.

John Calvin

Calvin rejects previously given explanations for the question of the four rivers.[9] He rejects the possibility that Pishon and

---

[9] Calvin, *Calvin's Commentary on the Bible.*

Gihon refer to the Ganges, the Nile, or even the Danube, because these are far from the Tigris and the Euphrates. He also rejects the possibility that the flood changed the surface of the earth sufficiently to move the springs of the Tigris and Euphrates. He insists that even after the flood, it was still the same earth and would not have had such a radically altered geography.

He argues that Moses accommodated his topography to the capacity of his age. What he means is that Moses did not give a scientifically precise description of the rivers, but one suited to the ability of his audience. He compares this to the issue of the moon, which is described as a great light even though it is smaller than the planets. The Bible is describing the world as it appears by common observation.

Calvin's proposal is based on the fact that the Tigris and Euphrates come close to each other at a point near the ancient city of Babylon. Furthermore, water travels from the Euphrates into the Tigris by various channels, canals, and ditches. Downstream, the two rivers diverged and headed in different directions before ending separately in the sea. Because they are connected, it is possible to reach every part of the Tigris-Euphrates river system while travelling along the water.

Calvin suggests that Eden was located in the region where these rivers come close together. Upstream of Eden, the rivers are simply the Tigris and the Euphrates. However, downstream of Eden, the Tigris becomes the Pishon while the Euphrates becomes the Gihon. Thus, in his scheme, the Tigris and Euphrates flow into Eden while the Pishon and Gihon flow out of it. Within Eden, they are all interconnected. This, according to Calvin, is what the text is actually describing.

However, the text describes the river as flowing, or going forth, and dividing into four rivers. This would be, at the very best, a highly unusual way of describing the geography that

Calvin proposes. Calvin argues that when the text indicates that the water went forth, it simply means that it was flowing water without intending to indicate the direction of the flow. As such, the water in some cases is flowing out of Eden and, in other cases, into Eden.

Calvin argues that the land around Babylon, where these rivers approach each other, is highly fertile and thus a plausible location for the Garden of Eden. Furthermore, he cites a couple of verses that refer to a land of Eden in these parts:

> Have the gods of the nations delivered them, the nations that my fathers destroyed, Gozan, Haran, Rezeph, and the people of Eden who were in Telassar? (Isaiah 37:12 ESV)

> Haran, Canneh, Eden, traders of Sheba, Asshur, and Chilmad traded with you. (Ezekiel 27:23 ESV)

The Hebrew word translated as Eden here is related to, but not the same word as, the Eden of the Garden of Eden. Most would take it as referring to another place with a similar name, not the location of the garden.

His proposal also requires that the rivers have different names upstream and downstream of Eden. As support, he cites Pliny, who records that the river Tigris is called Diglito near its source.[10] However, Calvin goes on to claim that "but after it has formed many channels and again coalesces, it takes the name of Pasitigris. But this is not mentioned by Pliny. Indeed, the Pasitigris is another river in the region, not the same thing as the Tigris.

Calvin cites ancient texts, which state that Pasitigris was called Pasin by its inhabitants. The name Pasin is somewhat similar to the Biblical term Pishon. Calvin then suggests that

---

[10] Pliny, *The Natural History*, Book 6, Chapter 31.

the name Pishon became Pasin, which became Pasitigris, which is now known simply as the Tigris.

Calvin argues that the Biblical description of Pishon as surrounding Havilah makes sense because the Tigris has a winding course below Mesoptomia. He cites a brief reference to Havilah later in Genesis in the description of where Ishmael's descendants settled:

> They settled from Havilah to Shur, which is opposite Egypt in the direction of Assyria. He settled over against all his kinsmen. (Genesis 25:18 ESV)

Since Shur is stated to be next to Egypt, Calvin takes Havilah to be on the opposite side of the Arabian peninsula, where the end of the Tigris was found. Furthermore, he suggests that the description of gold and precious stones is very applicable to this area.

In Calvin's proposal, the Gihon refers to the lower portion of the Euphrates. This poses a difficulty since that river is described as running through Cush, typically taken to refer to Ethiopia. Calvin argues that the land of the Midianites is also called Cush by appealing to the story of Moses marrying a Cushite woman:

> Miriam and Aaron spoke against Moses because of the Cushite woman whom he had married, for he had married a Cushite woman. (Numbers 12:1 ESV)

Moses had previously married Zipporah when he lived in Midian:

> And Moses was content to dwell with the man, and he gave Moses his daughter Zipporah. (Exodus 2:21 ESV)

Calvin's argument is based on the assumption that Zipporah is the Cushite woman, and thus the name Cush can also refer to her homeland, Midian. However, it is also possible that

Moses either took a second wife or that Zipporah had died and Moses got remarried. Indeed, the story of the Cushite wife makes very little sense if it is Zipporah. Why would Aaron and Miriam suddenly have a problem with Zipporah after years of her being married to Moses?

Calvin proposed a unique solution, interpreting that text as referring to known geography. However, his exegesis is suspect. In the primary passage, the geography that he points to would not typically be described as water flowing out of Eden and dividing into four rivers. He attempts to bolster his case with brief references to regions in other passages. But in each case, it is not clear that we should draw the same conclusion he does from that passage.

Conclusions

Historical Christians

Up to the Reformation, the answers given about this geography are fairly consistent. The Tigris and Euphrates are the same rivers that we know today. The Pishon is identified as the Ganges (except for Ephrem, who identifies it as the Danube), while the Gihon is the Nile. The authors sought to reconcile the text with the fact that these rivers clearly have distinct sources by claiming that the rivers flow underground from paradise to their apparent sources. Martin Luther accepted the traditional identification of the rivers. However, rather than postulate waters making their way from paradise, he suggests the flood was responsible for altering the geography. John Calvin reinterpreted the text as referring to known geography by not taking it as a description of a river flowing into four other rivers, but rather two rivers flowing into Eden and two other rivers flowing out.

Modern Science

Modern science has confirmed that the rivers traditionally identified as those in Genesis 2 have distinct sources. Furthermore, they are clearly not connected by any kind of underground waterway. The situation is as it appeared: these are distinct rivers with distinct sources.

Furthermore, river bifurcation is rare. Rivers tend to join together into larger rivers as they travel to the sea. They do not generally split into multiple rivers. Rivers will sometimes bifurcate, but the bifurcation tends to be unstable. Over time, rivers will favor one side or the other, digging out that side to carry all the water and leaving the other side dry. Exceptions are found in cases where rivers flow through exceptionally hard material that the water cannot dig out. Consequently, the four-way bifurcation described in the text is unlike anything we find anywhere on earth.

Additionally, water does not make its way from the ocean into rivers through springs. As discussed in the previous chapter, the ancients considered springs receiving water underground from the ocean to be the major source of water in rivers rather than rain. Against that background, it seemed plausible that these four different rivers could be fed by a single underground spring. However, it is now clear that this is not how rivers are fed.

Evaluation of the Response

Prior to the Reformation, there was general agreement about the identification of the rivers: they were the Tigris, the Euphrates, the Ganges, and the Nile. There was also general agreement that they shared an underground source of water coming from paradise. Neither idea has much support today. I'm unaware of anyone today who would defend either claim.

Neither idea comes from the Bible. The traditional identification of the rivers Pishon and Gishon has little to recommend it Biblically. There is nothing in the Biblical text that suggests that water flows underground from paradise to these four rivers. Both of these are extra-Biblical speculation.

Furthermore, this speculation was influenced by the science of their day. Historical authors tended to identify the rivers mentioned in this passage as being the major rivers in their conception of the world, such as the Danube, the Ganges, or the Nile. They already thought that rivers were fed from underground springs coming from the ocean, so the idea that these four rivers could be fed from paradise underground seemed plausible.

This issue provides another demonstration of historical respect for the authority of Scripture. All of these authors insisted on believing what the Scriptures were teaching about these rivers. They were not willing to allegorize the account. Even when the text seemed at odds with the facts of geography, they insisted that the text was accurate.

Today, there is no generally accepted solution to this difficulty. Modern young-earth creationists hold that the earth was so entirely reshaped by the flood that pre-flood geography bears no resemblance to modern geography. This explains any discrepancies, but it makes it difficult to explain why there are so many names from ancient Near East geography in the text. Some interpret the text as describing the coming together of four rivers in what is now the Persian Gulf.[11] But it is difficult to defend reading the Biblical text as referring to a confluence of rivers. Others argue the passage should simply not be taken a historically or geographically accurate.

---

[11] "The Garden of Eden."

Both approaches face difficulties because they are not going back to the source. They are both attempting to shoehorn the text into a particular paradigm. They are both consequently forced into exegetical awkwardness in an attempt to justify their interpretation. The solution, I would suggest, is that we need to go back to the source and carefully evaluate what this text is saying. Perhaps my suggestion that that original river was scattered around the ancient world is worth consideration. Perhaps there is a better explanation. Whatever the correct explanation, we need to go back to the source to find it.

Chapter 16

Was the Sun Created in the Middle of Creation Week?

The Question

The Genesis 1 creation account describes the creation of the sun, moon, and stars:

> And God said, "Let there be lights in the expanse of the heavens to separate the day from the night. And let them be for signs and for seasons, and for days and years, and let them be lights in the expanse of the heavens to give light upon the earth." And it was so. And God made the two great lights— the greater light to rule the day and the lesser light to rule the night—and the stars. And God set them in the expanse of the heavens to give light on the earth, to rule over the day and over the night, and to separate the light from the darkness. And God saw that it was good. And there was evening and there was morning, the fourth day. (Genesis 1:14–19 ESV)

However, the creation of the sun in the middle of Creation Week raises a number of questions. What was the source of the light created on the first day, if not the sun? How could there be mornings and evenings on the preceding three days if the sun had not yet been created? This is especially true on the first day, when there is neither a sun nor the heavens. Furthermore, how did the plants created on the third day grow if there was no sun?

One possibility is that the author never intended his account to be taken as a straightforward historical account of the process of creation. Creation Week may simply be some sort of literary device. If that were the case, he would have no reason to worry about whether or not the account made sense. Pointing out this difficulty is the basis for a long-standing argument for not taking the Genesis 1 passage as an historical account.

Other interpretations take the days of Genesis 1 to be an historical account, but take the days as something other than ordinary days. A substantial difficulty with such an interpretation is that the text explicitly identifies day and night with the light, darkness, morning, and evening:

> And God said, "Let there be light," and there was light. And God saw that the light was good. And God separated the light from the darkness. God called the light Day, and the darkness he called Night. And there was evening and there was morning, the first day. (Genesis 1:3–5 ESV)

Day is identified as the name for the light, which seems to make it clear that the light referenced here is daylight. The name night is given to the darkness opposed to the day. Furthermore, the day is described as having both a morning and an evening. It is difficult to see how the text could be referring to anything other than an ordinary day.

On the other hand, those who take the days of Genesis 1 as something other than ordinary days often point out that another verse in the early chapters of Genesis uses day to mean something other than an ordinary day:

> These are the generations of the heavens and the earth when
> they were created, in the day that the LORD God made the
> earth and the heavens. (Genesis 2:4 ESV)

The same Hebrew word is used for day in this verse as the days of Creation Week. However, in this case, the verse is referring not to an individual day but to the entire span of creation. The same idiom is also used in English. Consider phrases like "the day of the dinosaurs," "the day of Queen Elizabeth," or "my father's day." Similar language was used by Winston Churchill when he referred to Britain's "finest hour." These phrases do not refer to literal days or hours, but more generically to a period of time. But there seems no clear reason to think that this idiom is operating for the days of Creation Week.

For those who take the days of Genesis 1 as ordinary days, the implication is that there must have been another light source prior to the sun. But why would God create this transitory light source rather than just creating the sun? What happened to that light source? How did it happen that this other light source caused day and night just as the sun would?

A common idea historically was that God created some sort of light source, and then later created the sun, moon, and stars out of that light source. According to this idea, God created an original light on the first day. This light was divided up into the numerous heavenly bodies that we know of today on the fourth day.

This theory has a certain elegance to it. Scripturally, the text is probably vague enough to allow for this possibility. It fits the theme that God uses previously created materials during Creation Week. The heavens and the seas were created from the primordial waters of creation. The earth was created out of the sea. Animals were created from the earth. There is a certain plausible symmetry to the idea that the heavenly bodies we know of were also created from prior material.

However, the text does not actually explicitly state that heavenly bodies were created from the prior light. At best, we can suggest that the author intended to implicitly convey this idea. It is arguably unexpected that the author would not explicitly state that the heavenly bodies were created from the original light since the author does explicitly make such statements about the origin of other materials.

Another idea, historically, was that the day had a light of its own apart from the sun. One reason to think this was that there is light before the sun rises and after it sets. Today, we understand that this is the sun's light, visible because the sun is just around the curvature of the earth. However, this was not clear to those in the ancient world. As such, some thought the day had a light of its own; the sun merely added to that light.

If this theory were correct, it would make sense if the light created on the first day was daylight. The sun, which adds its own light to the day, was created on the fourth day. But since, according to this idea, daylight is distinct from sunlight, this poses no difficulty. Nevertheless, we have no Biblical indication that the ancient Israelites conceived of daylight as being distinct from sunlight.

The Scientific Discussion

A common view in the ancient world, promoted by people such as Aristotle, was that the heavens were eternal and immutable. The heavens had existed in exactly the same form for all eternity. A consequence of this view was that the sun had always existed. It was not brought into existence at any point, let alone after the earth.

According to classical science, the earth was at rest in the center, or bottom, of the universe. The entire heavens rotated around the earth. The sun's revolution around the earth caused day and night. If there was some other light created before the sun, did it also revolve around the earth? If the heavens did not exist until the second day, how could the sun partake in that daily revolution?

Historical Christians

Theophilus of Antioch

Theophilus mentions this issue in his description of the creation days.[1] He writes that God knew that the "vain philosophers" would say that plants were produced by the heavenly bodies, "so as to exclude God." But God created the plants and seeds before the sun, and thus the sun cannot be their cause.

Origen of Alexandria

Origen invokes this issue as an example of a statement that must be read only allegorically or spiritually instead of as a literal historical statement:[2]

> Now who is there, pray, possessed of understanding, that will regard the statement as appropriate, that the first day, and the second, and the third, in which also both evening and morning are mentioned, existed without sun, and moon, and stars — the first day even without a sky?

For Origen, and historical Christians generally, all of Scripture contains hidden allegorical meanings. Even when a passage describes a real historical event, there is still a hidden allegory conveying spiritual truths hidden in the text. Origen argues that in some passages, only the hidden spiritual meaning is

---

[1] Theophilus of Antioch, "To Autolycus," Book II, Chapter 15.
[2] Origen, Cavadini, and Lubac, *On First Principles*, Book 4, Section 16.

actually true; the historical events described in the passages either did not or could not have taken place. In fact, Origen argues that the Holy Spirit inspired the authors to put obviously impossible statements in the text so that the reader would know to look beyond the surface-level meaning of the text for the deeper allegory. The problem of explaining how the sun could have been created in the middle of Creation Week is Origen's first example of such an obviously impossible Biblical statement.

In another book, Origen quoted the pagan critic Celsus, who made this objection:[3]

> By far the most silly thing is the distribution of the creation of the world over certain days, before days existed: for, as the heaven was not yet created, nor the foundation of the earth yet laid, nor the sun yet revolving, how could there be days?

Origen responded[4] by suggesting that Celsus had not understood the days of creation correctly. He quotes the creation account:

> These are the generations of the heavens and the earth when they were created, in the day that the LORD God made the earth and the heavens. (Genesis 2:4 ESV)

The text does not refer to the days of creation but to the single day. Origen pointed out that Moses had surely not forgotten that he had just stated that the world had been created over the course of six days, not a single day. Therefore, Moses must have intended a special meaning.

Origen does not go into detail about the correct understanding of the creation days because he had already done that in his commentary on Genesis. Unfortunately, his commentary on Genesis is now lost. What we do know is that he "found

[3] Origen, "Contra Celsum," Book 6, Chapter 60.
[4] Origen, Book 6, Chapter 50.

fault with those who, taking the words in their apparent sig-nification, said that the time of six days was occupied in the creation of the world."[5]

Ephrem the Syrian

Ephrem discussed the issue in his description of creation.[6] In his account, the first day begins at evening. At this time, the earth was a watery mass surrounded by clouds. These clouds prevent light from God's supercelestial realm from reaching the earth, thus leaving the earth in darkness. At dawn, God creates the original light, which brings light to a previously dark earth.

Ephrem describes the light as being like a bright mist. Its light might have resembled either dawn or the fiery pillar that guided the children of Israel. The light spread around every-thing, chasing away all darkness. The light remained until the end of the day. That evening, God divided the waters above and the waters below, and the light into the waters below.

Ephrem states that this original light played a role in "the conception and the birth" of the plants created on the third day. However, it is left to the sun to ripen those plants when it is created. He reports that "it is said" that the sun, moon, and stars were then created from this original light on the fourth day.

Basil of Caesarea

Basil states that God created a primitive light on the first day, which spread abroad or was withdrawn to produce day and night.[7] When dry land appears on the third day, Basil says that

---

[5] Origen, Book 6, Chapter 60.

[6] Ephrem the Syrian, *Commentary on Genesis*, Chapter 1.

[7] Basil of Caesarea, "Hexaemeron," Homily 2, Section 8.

God was showing that we should not attribute the drying of the land to the sun because the sun was not yet created.[8] He states that God did not create the sun until the fourth day so that those who lived in ignorance of God would not consider the sun as the "origin and father of light, or the maker of all that grows out of the earth."[9]

However, if the light had already been created, why does the Scripture say that the sun was created in order to give light to the earth? Basil argues that the nature of light was created on the first day, but the sun was created as a vehicle for that light on the fourth day. He compares it to a lamp, which is not fire but can be a vehicle for fire.

Basil appeals to the story of Moses and the burning bush to support the possibility of the light's existing apart from the source of the light. In that story, God created a fire that was bright but did not consume the bush. Basil argues that a similar division will take place in the afterlife, with the wicked feeling the heat of fire, but not the light, while the righteous see the light of fire but do not feel its heat. If God can separate the burning from the fire, he could also separate the light from the sun and create them separately.

Basil also appeals to the phases of the moon in order to further support his idea. At different phases, the moon takes on different shapes. However, the moon is not actually changing shape but rather is only partially lit up. This again demonstrates the separability of the light from the vehicle of the light.

---

[8] Basil of Caesarea, Homily 4, Section 5.
[9] Basil of Caesarea, Homily 6, Section 2.

Ambrose of Milan

Ambrose argues that the light of the sun is distinct from the light of day.[10] He argues that this can be seen by the fact that there is light before sunrise and after sunset. The sun only makes the day brighter, making the day brightest at noon when the sun is directly overhead.

He also argues that the sun provides heat, but the light visible before the sun rises and after the sun sets does not, showing it to have a different source of light. He compares this to Moses's burning bush. That fire had the power of light but not that of consuming the bush. Likewise, daylight has the power of light but not heat. However, the sun has the power of both light and heat.

Later, Ambrose addresses the question of the growth of crops before the creation of the sun.[11] He notes that many say that the earth could not have germinated without the warmth of the sun's heat. Accordingly, many had "bestowed divine honors on the sun." God makes it clear that this is not the case by making the plants grow before the sun. It is God who gives plants life, not the sun.

John Chrysostom

Chrysostom discusses this question while discussing the creation of the sun.[12] He points out that sun worship is very common among the pagans. Foreseeing this, the Scripture makes it clear that the sun was not created until after the plants. Thus, nobody can say that plants need the sun in order to be brought forth from the earth. We must instead attribute them to the Creator.

---

[10] Ambrose of Milan, "Hexameron," Chapter 3, Sections 7-9.

[11] Ambrose of Milan, Chapter 6, Section 27.

[12] Chrysostom, *Homilies on Genesis*, Homily 76, Section 12.

He clarifies that he does not reject the idea that the sun makes a contribution to the ripening of crops. He compares the sun to a farmer, who clearly contributes to the growth of crops. Nevertheless, even the work of thousands of farmers would be worthless unless God initiated the process. Likewise, even the work of farmer, sun, moon, and climate combined, would have no effect unless God played his role. These other factors make their efficacious contribution only if God's mighty hand is ready.

Chrysostom goes on to argue that God created the sun on the fourth day, lest we think the sun causes day.[13] Clearly, if there were days before the sun, the sun is not itself the cause of the day. Rather, the Lord only uses the sun to make the day brighter or more brilliant than it would otherwise be.

Severian of Gabala

Severian gives a number of reasons for the creation of the sun on day four.[14] He states that it is done to combat polytheism. It was also not done prior to day two because there was not yet a sky in which to put the sun. It was not done prior to day three because there were not yet any plants that needed sunlight. It was also delayed to make it clear that plants were not produced naturally by the sun. He explains that the sun was created out of the light made on day one, comparing it to the division of the mass of waters in seas, lakes, and rivers.

Augustine of Hippo

Augustine defends against the objection made by the Manichees that there could not be days before the creation

---

[13] Chrysostom, Homily 76, Section 14.
[14] Severian, "Homilies on Creation and Fall," Homily 3.

of the sun.[15] His first response is to point out that we can sense the passage of time without the sun. Even if we were in caves, without the ability to see the sun, we could sense that three days worth of time had passed. Nevertheless, Augustine quickly admits that this answer will not work because the text speaks of morning and evening, which require not merely the passage of time but the sun. Augustine suggests that this refers to God's starting and stopping of his work, just as men tend to work from morning until evening. He points out that Scripture often uses human language to describe God.

In his later works, *On the Literal Meaning of Genesis* and *The City of God*, Augustine interacts with the issue again as part of his larger argument that the light created on Day One is not a material light but instead a spiritual light associated with angelic beings. Augustine is influenced here by Neoplatonic thought, in which souls and spiritual beings are imbued with light deriving from the One.

In order to support this interpretation, Augustine is forced into some rather awkward readings of the text. He explains day, night, evening, and morning in terms of the knowledge of angels. He thinks that the days of Genesis 1 are not a sequence of days but rather events that all took place simultaneously. On this point, he points to several verses that he thinks support the idea that creation happened all at once. However, all of his seeming support from those verses is due to flaws in the Latin translation Augustine was using.

However, our concern here is the particular issue of the sun's being created on Day Four. This difficulty is a major argument that Augustine makes for his position. Each time that he runs into difficulty expositing the text in line with his idea that the light is spiritual, he refers back to this argument to

---

[15] Augustine, "Two Books on Genesis against the Manichees," Book 1, Chapter 14, Section 20.

bolster his position. In an oft-quoted passage discussed in the preface where Augustine warns against taking interpretations that might conflict with science and reason,[16] he probably has this issue in mind.

Augustine acknowledges that the text seems to describe a sequence of days, nights, evenings, and mornings.[17] He argues that God could not reasonably have taken the whole day to create, divide, and name the light and the darkness. However, he acknowledges that God could have created the light instantaneously, with the light remaining in place for the course of the day. Nevertheless, Augustine points out that it is always day somewhere on earth. The only way it makes sense to speak of day and night is from the vantage point of a particular place on earth. This makes no sense from God's vantage point, but it could make sense from the vantage point of the place where God would later create man.

Augustine asks why God make the sun if the original light was sufficient to make day and was even called day?[18] He mentions three theories about this: the light was too far away to shine on the earth, the sun made the day brighter than the original light, and the sun was made out of the original light.

All of these theories leave unanswered the question of what happened to the original light in the evening. Augustine does "not think any explanation can be found."[19] He simply rejects the idea that the light could have been extinguished each evening and rekindled each morning. Nevertheless, there remains the obvious answer: the primitive light revolved around the earth, providing daylight to one side while the other side experienced night. In response, Augustine spends chapters

---

[16] Augustine, *The Literal Meaning of Genesis. 1*, Book 1, Chapter 19.
[17] Augustine, Book 1, Chapter 10.
[18] Augustine, Book 1, Chapter 11.
[19] Augustine, Book 1, Chapter 11.

12 through 15 arguing that the primitive earth was formless matter and thus would not have blocked the light, producing day and night on opposite sides of the earth.

Augustine also argues against the idea that the light was diffused to produce the day and contracted or drawn back up to produce the night.[20] There were no creatures living on earth who could have benefited from the day-night cycle. Moreover, light simply does not work this way. He acknowledges that, according to the science of his own day, there was a ray of light coming from human eyes that was pulled in and out to look at distant or near objects. However, he argues that this light is very weak and indistinguishable from outside light and thus does not demonstrate light's ability to be diffused and contracted.

Augustine objects to all previously given explanations. This is part of his larger argument, wherein he attempts to argue that the light created in Genesis 1 is a spiritual light associated with angels.

Thomas Aquinas

Aquinas considers two related questions "whether the production of light is fittingly assigned to the first day"[21] and "[w]hether the lights ought to have been produced on the fourth day?"[22]

Aquinas considers the possibility that the original light was some sort of "luminous nebula." He rejects the possibility that there could be some light that has since ceased to exist, insisting that the creation account describes the creation of nature as we know it. He rejects the possibility that the nebula

---

[20] Augustine, Book 1, Chapter 16.

[21] Aquinas, *The Summa Theologiæ of St. Thomas Aquinas*, The First Part, Question 67, Article 4.

[22] Aquinas, The First Part, Question 70, Article 1.

still existed, but was so near the sun as to be indistinguishable from it. He rejects this as "superfluous" insisting that God made nothing in vain. He also rejects the idea that the sun was made from this primitive light. Aquinas points out that if the sun is made of immutable aether, it is not capable of having previously taken on a different form because being immutable, it cannot have changed.

Aquinas offers Augustine's theory of instantaneous creation as a possible answer. However, his primary answer is that the primitive light was from a primitive version of the sun. This primitive heavenly body consisted of what would later be the substance of the sun and possessed light but was not yet fully formed. This is similar to the idea of the sun's being created from primitive light, but it maintains consistency with the immutable aether of classical physics by maintaining that the substance and form of the sun remained the same. But why is this primitive light not mentioned explicitly in the text? Aquinas suggests it was due to a need to guard people from idolatry.

Aquinas considers the argument that day and night are caused by the daily revolution of the heavens, but the heavens were only created on the second day.[23] As such, it does not make sense for the light of day to have been created on the first day. He considers the proposal of Basil that the light was expanded and contracted to produce the first days and nights. He rejects this as unnecessary since no living creature could benefit from the light, as well as being contrary to the known nature of light. He admits that God could miraculously expand and contract the light, but insists that we ought not to look for miracles in creation.

---

[23] Aquinas, The First Part, Question 67, Article 4.

He responds by distinguishing between the general movement of the heavens and the special movement of the individual stars. He states that the general movement of the heavens, which is responsible for the day-night cycle, began on the first day rather than the second, whereas the special movements began on the fourth day. Aquinas does not explain how the heavens revolved before they were created.

Bede

Bede takes there to be a primitive light that existed prior to the sun in his commentary on Genesis 1:5.[24] He considers two possible forms this light could take. The light could revolve around the earth like the sun. Alternatively, it could be that this light gradually grew brighter in the morning and then gradually grew darker at night.

Bede argues that the text rules out the possibility of a gradually increasing or decreasing light, thus showing that the original light revolved around the earth. The text speaks of each day as evening and morning, counting each day from one evening to the next. This makes sense if the sun is revolving around the earth, because a day can be counted as one complete revolution around the earth from evening to evening. However, if the light appeared and disappeared, it would only make sense to count days as being from the time of the beginning of the light to its extinguishing. In that case, it would only make sense to count days from morning to evening and not from one evening to the next.

Martin Luther

Luther discusses the nature of the light that illuminated the earth before the creation of the heavenly bodies in his commen-

---

[24] Bede, "Commentary on Genesis."

tary on Genesis.[25] He dismisses any allegorical interpretation as absurdities which are not to be tolerated. It was a real material light. When it came to the question of whether or not this light moved with a circular motion around the earth, Luther confesses to not knowing the answer. Nevertheless, he gives as his opinion that the light moved around the earth like the sun, producing day and night. He notes the difficulty in understanding the nature of this light. Regardless, he insists on following the plain teaching of the text.

When commenting on Genesis 1:14, Luther addresses the question of what happened to the light. Did it disappear, did it become the heavenly bodies, or does it still persist? Luther notes that a wide variety of opinions have been offered on this question. He simply says that just as God first made the formless earth and later formed and adorned it, he created the primitive light and later perfected and completed that light by the creation of the heavenly bodies.

John Calvin

In his commentary on Genesis,[26] Calvin declares that God made the plants before the sun so that we may learn that he is the original of all things. God shows that he does not need the sun but can cause plants to grow without its assistance.

Conclusions

Historical Christians

The historical Christians we have looked at nearly universally believed in some prior light that served the purpose of the sun before the sun was made. The two explicit exceptions are

---

[25] Luther, *Luther On The Creation: A Critical and Devotional Commentry on Genesis.*
[26] Calvin, *Calvin's Commentary on the Bible.*

Origen and Augustine. Origen takes the text as exclusively allegorical. Augustine gives what he would insist is a literal reading in which the light refers to angelic beings.

Basil, Severian, and Aquinas all give accounts in which the light somehow becomes the heavenly bodies. However, the details differ in each case. Basil speaks of the nature of light's being created on the first day, which is then placed into the body of the son. Severian states that the light was divided up to produce the heavenly bodies. Aquinas argues that the sun existed in some primitive form from the start of creation.

Chrysostom and Ambrose both espouse the idea that there is daylight separate from the sun's light. Ambrose is explicit on this point, offering arguments in an attempt to support it. Neither, however, seems to present the idea as something they expect to be controversial, suggesting that they are operating on what they take to be the science of their day.

A recurring theme is that God created the sun after plants in order to make it clear that he is the creator of plants and to combat the errors of polytheism and idolatry. The authors insist that while the sun plays some sort of role in the growth of plants, it is really God who gives plants growth.

Modern Science

The ancient idea of an immutable heaven, including a sun that had existed for all eternity, was rejected by modern science. Indeed, it is clear the sun is burning up and has a limited lifespan. In fact, modern scientists concluded that the earth could not be eternal precisely because the sun had a limited lifespan.[27] The sun neither has always existed nor will exist forever.

---

[27] Cajori, "The Age of the Sun and the Earth."

However, while the sun is clearly not eternal, modern science has not confirmed the traditional understanding of the Biblical creation account. The standard theory today for the origin of the solar system is called the nebular hypothesis. According to that theory, the sun originated hundreds of millions of years before the earth.

While classical science thought that the heavens were turning around the earth, modern science holds that the earth itself is rotating. This makes it easier to explain how a light source prior to the sun could have produced day and night. Any light source directed at the rotating earth would produce day and night. As such, it is less complex to explain how a previous sort of light could have functioned in modern science than it was in classical science.

Evaluation of the Response

The interpretation of Creation Week is one of the most controversial debates within modern Christianity. The modern, mainstream scientific consensus about the history of the solar system is irreconcilable with the traditional reading of the creation account. The modern Christian must either reject mainstream science's position or take a very different reading of the Biblical text.

A minority of authors, namely Augustine and Origen, interpreted the creation account in a way other than as a sequence of twenty-four-hour days. This avoids the difficulty posed by the creation of the sun after the light. Indeed, Augustine and Origen are frequently cited as precedent by those defending something other than the twenty-four-hour, six-day interpretation of Genesis 1.

One might wonder if modern science has vindicated this minority view. It is unclear exactly what Origen believed. However, Augustine's view has clearly not been vindicated. He held that the universe was created simultaneously and in an instant. The creation of the universe in an instant could not be more different from the gradual formation of the universe over billions of years.

Furthermore, it is difficult to argue that either of these authors was proposing the best interpretation of Scripture. Origen is looking for the hidden meaning in the text and deliberately disregarding what we now think of as authorial intent. Augustine is influenced by the philosophies of his day and based his conclusions on flawed translations. On the other hand, we cannot identify any flawed translations, clearly bad exegesis, or philosophical influence behind the other authors' reading of the days as a sequence of ordinary twenty-four days.

Ambrose and Chrysostom espoused the idea that there was daylight distinct from sunlight. This would turn out to be incorrect. However, this appears to have been a case of their espousing what they took to be a standard view in their time.

We see another example of respect for the authority of the Biblical text. Historical authors insisted on believing the Biblical text. Most thought the sun really was created during the middle of the Creation Week. Origen rejects this, probably taking some sort of allegorical interpretation. In this, he provides one of the few examples we have found of attempting to resolve a scientific issue by interpreting the text symbolically. Augustine gives a unique interpretation, but still insists on holding the text to be true and refuses to interpret it merely allegorically. Whether they interpreted the text as ordinary days or not, they believed what the text says.

Today, the problem is unsolved. There is no generally accepted resolution to this question. I would suggest that the solution must be to go back to the source. This could be to return the source of the scientific data and argue that it does not actually support the mainstream view. But this would require the science being well done and not merely shoehorning the data into an understanding that does not fit. It could also be to return to the Biblical source and show that it does not say what it has traditionally been taken to say. But this would require the exegesis to be done well and not merely attempt to shoehorn the text into something it does not say. Only an approach that does not shoehorn either the science or the Biblical text but truly goes back to the source will be successful.

Chapter 17

How Large Are the Heavenly Bodies?

The Scientific Issue

Is the sun larger than the moon? If we only consider the appearance of these heavenly bodies in the sky, we might naively conclude that they are approximately the same size. Furthermore, they appear to be no larger than a cloud. We might also conclude that the stars and planets are a tiny specks in comparison to either the sun or the moon. Alternatively, it is possible that these heavenly bodies are actually very large but only appear small in the sky because they are very far away.

One reason to think that they are very far away is that the heavenly bodies appear in a consistent place and with a consistent size, regardless of the observer's location on earth. Contrast this with a large earthly object, such as a mountain. The mountain will appear larger or smaller, depending on how close it is. It will appear in different directions, depending on one's location. On the other hand, one could travel the whole breadth of the ancient world, and the sun, moon, stars, and planets would all appear approximately the same size and position in the sky. Why? Because the heavenly bodies are so far away, any change in one's location on earth makes essentially no difference.

The sun can be seen to be larger and farther away than the moon because, during a solar eclipse, the moon passes in front of the sun and not the other way around. Furthermore, by observing the phases of the moon, we can see that the light of the sun hits the moon and the earth at a similar angle. In order for this to be the case, the distance between the earth and the moon must be very small compared to the distance between the earth and the sun. The sun must then be much larger than the moon.

There were also indications that at least some of the planets were actually larger than the moon. It takes much longer for the planets to move in their apparent circle around the earth than for the moon. For example, it takes the moon twenty-nine and a half days to orbit the earth, but it takes Saturn twenty-nine and a half years. According to classical science, the reason it takes Saturn so long to orbit the earth is because it is far away and thus has to make a much larger circle. Since Saturn appeared to take approximately 365 times as long to move around the earth as the moon, it was thought to be about 365 times as far away as the moon. This would easily make it much larger than the moon.

The stars were thought to be even farther away than the planets. As such, it was thought that they were much larger than they appeared. Indeed, they could be much larger than either the sun or the moon.

The Biblical Discussion

The Bible describes the creation of the sun, moon, and stars:

> And God made the two great lights—the greater light to rule the day and the lesser light to rule the night—and the stars. (Genesis 1:16 ESV)

A Psalm uses similar language to describe the lights:

> to him who made the great lights, for his steadfast love endures forever; the sun to rule over the day, for his steadfast love endures forever; the moon and stars to rule over the night, for his steadfast love endures forever; (Psalm 136:7–9 ESV)

What does it mean for the light to be great? The text indicates that the sun is greater than the moon, while both are considered great, while the stars are not described as great. But what does great mean in this context? The Hebrew word can be translated in a variety of ways, including: bigger, more important, older, higher, or louder. Generally, the sense of greatness is clear from what is described as great. A great voice is a loud voice. A great pit is a deep pit. The great priest is the chief or high priest. The great sea was the Mediterranean Sea, the largest nearby body of water. A great rain is a very heavy rainfall. Given how greatness applies to other objects, a great light is probably a very bright light. The author is probably referring to how much light we get from both the sun and the moon. In contrast, we get relatively little light from the stars; hence, they are not great.

However, greatness can mean large in size. It is possible to read the passage as referring to the sun and moon as the two large lights, with the sun being larger than the moon. However, if the author were referring to the apparent size of objects in the sky, he would not describe the sun as larger than the moon, because they have the same apparent size. It seems unlikely that the author was scientifically sophisticated enough to realize that the sun was actually much larger than the moon. Thus, he was probably not referring to the actual sizes either. Given the way the word great is used in the Old Testament, it seems most likely that the author was simply referring to how much light came from these objects.

The Apostle Paul briefly refers to the heavenly bodies differing in glory:

> There are heavenly bodies and earthly bodies, but the glory of the heavenly is of one kind, and the glory of the earthly is of another. There is one glory of the sun, and another glory of the moon, and another glory of the stars; for star differs from star in glory. (1 Corinthians 15:40–41 ESV)

In context, Paul is contrasting our current earthly bodies with the heavenly bodies we will receive at the resurrection. His point is that we will go from the limited glory of an earthly body to the greater glory of a heavenly body. Glory is a very broad concept. But it seems likely that Paul sees the heavenly bodies as differing in glory primarily due to their differences in brightness.

Historical Christians

Basil of Caesarea

Basil takes the greatness as a reference to the sizes of the heavenly bodies.[1] He asks whether the sun and moon are great or large in an absolute sense or merely relative to the stars. We might describe an ant as large, but only relative to other ants. The ant remains very small. Basil argues that the sun and moon are great without qualification because they are of a "prodigious size."

---

[1] Basil of Caesarea, "Hexaemeron," Homily 6, Section 9-11.

Basil warns his listeners not to think the sun is small just because it only takes up a small part of the sky. He points out that distant objects often seem small, appealing to the appearance of the plains below when looking out from a high mountain. Basil demonstrated the large size of the sun by pointing out that it always looks the same regardless of where the people watching are located or where in the heavens the sun and moon are located. Those in the east do not see the sun getting smaller as it sets, and those in the west do not see it getting smaller when it rises. Basil further points out that the myriad stars in the night sky are insufficient to light up the earth. However, the single sun is able to light up the whole earth due to its massive size. It is able to do this even before it has finished rising.

Basil goes on to make similar arguments, showing that the moon is also very large. He points out that the moon, like the sun, appears the same size to people in all places and at all times. At times, parts of the moon are not visible; however, the whole size remains the same. He also points to the believed effects of the moon on the weather, plants, and animals, arguing that these reflect the immense size of the moon.

Crucially, Basil believes that the testimony of Scripture is that the sun and moon are very large and not as small as they appear. He presents some scientific arguments to show that the Scriptures are correct on this point, thereby demonstrating their reliability.

Ambrose of Milan

Ambrose states that the sun and moon are great by their own right and not merely in comparison to other objects such as the heavens or the sea.[2] We must admit, he insists, that the sun is

---

[2] Ambrose of Milan, "Hexameron," Homily 4, Chapter 6.

mighty because it fills the whole earth with heat. Both the sun and moon can also be seen to be mighty because they fill the earth and atmosphere with light.

Ambrose appeals to the fact that the moon seems to be the same to all men to show that it is very large. Its light may change from time to time, but its appearance is the same for all men at any point in time. He points out that objects appear to change size as we move farther away from or closer to them, but the moon does not, showing how massive it is. The same is true of the sun. As an illustration, Ambrose claims, incorrectly, that the sun rises at the same instant in India as in Britain.

He tells his hearers not to be deceived by the fact that the sun appears very small in the sky. Our eyes can deceive us; therefore, we cannot trust them. If we see a plain from a mountain, or distant ships on the sea, they will seem much smaller than they actually are. He appeals to the sun's ability to fill the earth with warmth and light far more effectively than the myriad stars as further evidence, as well as the moon's believed effects on plants, animals, and weather.

Augustine of Hippo

Augustine noted that some had maintained that the stars are as large or perhaps even larger than the sun, but only appear small because of their great distance.[3] He responds by insisting that whatever the truth of this question, it is sufficient for Christians to acknowledge God as the creator of these heavenly bodies.

However, Augustine goes on to note that we must hold to the words of Paul speaking of the differing glories of the heavenly bodies. Augustine takes Paul as referring to the brightness of the stars, and considers the objection that the sun does not

---

[3] Augustine, *The Literal Meaning of Genesis. 1*, Book 2, Chapter 16.

in fact differ in glory from the other stars but rather is simply closer to us.

Augustine addresses this objection by arguing that the stars differ in glory from the perspective of men on earth. Paul, in context, is discussing the difference between our current bodies and our resurrection bodies. Augustine thus also suggests that our resurrection bodies will look different from their true nature, which is analogous to the way in which the stars may appear quite small even though they are very large.

Nevertheless, Augustine presents an objection to the idea that the stars are as large or larger than the sun. He argues that those who argued that the stars were larger than the sun and moon nevertheless attributed great significance to the sun. Why would the sun be so significant if the sun was actually smaller than many of the distant stars? Furthermore, some of the movement of what we know as the planets was attributed to the sun. But if these planets were larger than the sun, how could the sun have this effect? If these planets were smaller than the sun and other stars, why was such importance attached to their movement through the heavens?

Augustine states, "But let those who are strangers to our Father who is in heaven say what they will about the heavens." He insists that it is not necessary for us to engage in "subtle speculation" about the distances and sizes of the stars. Rather, we should simply believe, as the text indicates, that the sun and moon are greater than the stars, and the sun is greater than the moon. He says that it must be granted that this is true, at least as these heavenly bodies appear to us on earth:

> They will certainly grant this at least to our eyes, that these two lights obviously shine more brightly upon earth, that day is illumined only by the light of the sun, and that night with

all its stars does not shine as bright without the moon as when lighted by her rays.

Bede

In his Commentary on Genesis,[4] Bede wrote that the lights were not merely great relative to other things but were great "in terms of their own function." The sun is great because it is able to fill the whole earth with its heat, and the moon is great because it is able to fill the whole earth with its light. They illumine everything and are equally seen by all people. The fact that they appear the same size to all people is a "clear proof" of their great magnitude.

Bede indicates the sun is greater for a variety of reasons. It is more beautiful, provides more brilliant light, and provides heat. He notes that while the sun and moon appear to be of equal size, it is inferred that the sun is much farther away. He states that we cannot know exactly how large it is.

Remigius of Auxerre

In his commentary on Genesis,[5] Remigius wrote that the lights were called great not so much by comparison, but because they were of a great size. The sun gives life to the entire earth. The moon gives light but not heat to the same region. Part of this greatness is being seen equally by all people. Furthermore, the sun is great because it gives light to the moon. The moon and sun appear to be the same size, but the sun is much larger as it is much farther away from the earth.

---

[4] Bede, "Commentary on Genesis."
[5] Remigius of Auxerre, "Exposition on Genesis."

Martin Luther

In his commentary on Genesis,[6] Luther says the text has nothing to do with what philosophers say about the size of the heavenly bodies. Instead, it has to do with the magnitude of the light coming from those heavenly bodies. Luther states that even if all of the stars were collected together, the result might be larger than the sun, yet it would not form a light equal to the light of the sun. Furthermore, if the sun were divided into minute particles, even those particles would outshine the stars. Luther insists that there is an essential difference, citing Paul's statement that the heavenly bodies differ in glory.

John Calvin

In his commentary on Genesis,[7] Calvin wrote that the text is not discussing philosophically the size of the sun or the moon. Rather, it is speaking of how much light comes to us from them. Calvin explains that the moon is a lesser light because it emits less light than the sun. Moses is speaking to our senses and thus, to understand his meaning, we must understand him as speaking of how the light appears to us. Calvin indicates that the text rebukes those who would "censure Moses for not speaking with greater exactness."

Calvin points out that Moses does not mention the spheres that philosophers attributed to the heavens. Nor does he mention Saturn, the furthest known planet, which, even though it appears small, is actually larger than the moon. Calvin explains that Moses wrote in a popular style to be understood by all ordinary people. The astronomer investigates things that the common man does not need to understand. Calvin speaks highly of the merits of studying astronomy but insists

---

[6] Luther, *Luther On The Creation: A Critical and Devotional Commentry on Genesis.*

[7] Calvin, *Calvin's Commentary on the Bible.*

that Moses wrote instead in terms of appearances so that the common man may understand his words.

Some have taken this discourse as Calvin allowing Moses to be scientifically inaccurate because the Bible is not about astronomy. But Calvin actually argues that Moses is right, but he's not talking about the size of the sun, moon, or other heavenly bodies but the amount of light they emit to us. Calvin's point is one of subject, not accuracy. Moses speaks accurately, but he does not speak of astronomy.

Conclusions

Historical Authors

Several of the authors, Basil, Ambrose, Bede, and Remigius, emphasize how the sun and moon are very great. They put particular emphasis on how we know that the sun and moon are large objects, even though they only take up a small part of the sky. On the other hand, Augustine, Calvin, and Luther argue the text is speaking in terms of appearance, only indicating something about how much light the earth receives from these heavenly bodies. Augustine and Calvin both explicitly mention the idea that some of the stars or planets are larger than the sun or the moon. Luther and Augustine both mention Paul's statement about the differing glories of heavenly bodies. Augustine explains this as the language of appearance, but Luther takes it as describing a difference in the nature of the heavenly bodies.

Modern Science

Modern science has confirmed the general picture of the sizes of the heavenly bodies. The sun and moon are indeed very large compared to terrestrial objects. The sun is far larger than the moon, and the planets and stars are larger than the moon and. The stars are, in some cases, larger than the sun. The actual estimates of sizes obtained in the classical world would turn out to be not very accurate. This was mostly due to the imprecise measurements of the classical world, given that they lacked the technology to measure any aspect of astronomy precisely. The approximate size of the solar system was not determined until the 18th century.Indeed, the estimates of the distances in the solar system were still being refined during the twentieth century.

Evaluation of the Response

The authors who emphasized the great size of the sun and the moon ran into difficulty. If the Bible intends to indicate the massive actual size of the sun and moon, this is accurate. However, the stars and planets, which are not described as great lights, are also very large. The stars and planets are larger than the moon and stars are larger, in some cases, than the sun or the moon. On the other hand, those authors who argued that the text simply did not speak of the actual sizes of these objects but merely the amount of light they provided to the earth, would turn out to be correct.

None of the authors who interpreted the passage as referring to the great size of the sun and moon shows awareness of the issue of planets or stars being much larger than they appear. Two of the three authors who interpret the passage as not referring to their actual size reference the issue of these larger heavenly bodies. Luther is the one author who does not, although given how late Luther lived in history, he probably has the issue in mind.

Thus, we have a situation where Christians read scientific information into a text that the author would not plausibly have intended. It was easy to read the classical science ideas about the sizes of the sun and moon into the text. This later backfired as it became clear that, in the bigger picture, the science thought to be in the text is not quite right. The stars, not considered great lights in the text, are actually very large but much further away. The authors, who are aware of the issue, avoided it by properly considering what the Biblical author is talking about instead of trying to find scientific knowledge in the Scriptures. They went back to the source.

Even though their exegesis was problematic, we see that historical Christians thought that Biblical statements about the stars held authority on scientific questions. They treated the creation account and Paul's statement as making scientific claims that Christians had to hold to. They did not treat these passages as beyond the scope of Biblical authority nor did they interpret them as symbolic or allegorical.

There is also encouragement in the Church's handling of this issue. Yes, it is unfortunate that some took the Bible as making a claim that they thought to be true. But when forced to look more closely at the issue, it was quickly seen as clear that the Bible does not make claims about the actual sizes of these heavenly bodies. The problem was reading the Biblical text in the light of classical science. The solution was going back to the source.

Chapter 18

Does the Moon Have Its Own Light?

The Scientific Question

The moon appears to be a light in the sky. One might easily assume that it generates its own light. However, classical science concluded that the moon actually reflected the light of the sun rather than generating its own light.

A crucial piece of evidence for this conclusion was the changing shape of the moon. It goes from a new moon, to a crescent moon, to a half moon, to a full moon. This is explained by the idea that the moon is lit up by the sun. Only the half of the moon facing the sun would be lit up. When the lit-up side of the moon is facing away from the earth, the side facing the earth is entirely dark, which we call the new moon. When the lit-up side of the moon is facing the earth, the moon is full. In between, the lit-up side is partially facing the earth, producing crescent, half, or gibbous moons.

This also explains why the moon's light is so much less bright and warm than the sun's light. The moon's light is merely a pale reflection of the sun's light. The intense heat and light of the sun are understandably much reduced in the reflection of the moon.

The Biblical Discussion

The Bible contains a few statements that could be taken to indicate that the moon has its own light. The Genesis 1 creation account describes the creation of the moon:

> And God said, "Let there be lights in the expanse of the heavens to separate the day from the night. And let them be for signs and for seasons, and for days and years, and let them be lights in the expanse of the heavens to give light upon the earth." And it was so. And God made the two great lights—the greater light to rule the day and the lesser light to rule the night—and the stars. (Genesis 1:14–16 ESV)

This text describes the moon as a light and states that it gives light to the earth. But if the moon does not produce any light of its own, is it really a light?

In common parlance, the reflections of light are frequently referred to as lights. One example would be searchlights. Usually, when someone speaks of seeing searchlights, they do not mean the machine that produced the light; they mean the reflected light that they are seeing in the sky. If one were to observe a planet in the sky, one would describe it as a bright but small light, even though we know that planets are simply reflecting the light of the sun. When we observe bright parts of the sky, we call them lights.

This is because we use the term light to refer to sources of light even if they are not the original source of the light. We observe light coming to us from a particular place and we call that a light.

In order to argue that the text claims that the moon is the source of its own light, one would have to show that Hebrew is more specific in usage than English. However, Hebrew is usually much less specific and precise in its terminology. As such, it seems clear that the text simply does not indicate anything about the source of the moon's light.

Other verses refer to the light of the moon, for example:

> Moreover, the light of the moon will be as the light of the sun, and the light of the sun will be sevenfold, as the light of seven days, in the day when the LORD binds up the brokenness of his people, and heals the wounds inflicted by his blow. (Isaiah 30:26 ESV)

> Immediately after the tribulation of those days the sun will be darkened, and the moon will not give its light, and the stars will fall from heaven, and the powers of the heavens will be shaken. (Matthew 24:29 ESV)

Some might be inclined to argue that moonlight is actually the sun's light because the sun produces it. However, one would never argue that your house isn't your house because you bought it from someone else, or that your money isn't your money because somebody paid it to you, or that your dog isn't your dog because you didn't produce the dog. As such, the fact that the moon gets its light from the sun does not mean the light does not become the moon's.

Historical Christians

Origen of Alexandria

Origen refers to the moon's light's coming from the sun in an allegory of the Church, saying:[1]

> For just as the moon is said to receive light from the sun so that the night likewise can be illuminated by it, so also the

---

[1] Origen, *Homilies on Genesis and Exodus*, Homily 1, Section 5.

Church, when the light of Christ has been received, illuminates all those who live in the night of ignorance.

Nevertheless, he does not appear to be aware of any objection that might be raised against the account for incorrectly describing the moon as a light.

Basil of Caesarea

Basil refers to the light of the moon's coming from the sun.[2] In context, he is making the point that the moon is large and does not actually change size as it moves through its phases. He tells his hearers, "Do not tell me that the light of the moon is borrowed," insisting that this is beside the point of the subject he is discussing.

Gregory of Nyssa

Gregory of Nyssa refers to the moon's lacking its own light.[3] He points out that by observing the phases of the moon and the fact that the lit part of the moon is always facing the sun, we can discern that the moon's light does not come from the moon itself. His overall purpose is to argue that we have an intelligent, immortal soul distinct from our bodies that can take part in a future resurrection. Understanding the nature of the moon's light is an example of something only such an intelligent soul could discern.

Ambrose of Milan

Ambrose refers to the moon's borrowing its light from the sun.[4] He is making the point that the moon does not actually grow or shrink, but merely gives up its light. Ambrose uses this as

---

[2] Basil of Caesarea, "Hexaemeron," Homily 6.
[3] Gregory of Nyssa, "On the Soul and the Resurrection."
[4] Ambrose of Milan, "Hexameron," Homily Four, Chapter 2.

an analogy or allegory for the Church, saying that the Church does not lose its light, even if at times it seems diminished.

Augustine of Hippo

Augustine touches on this subject while discussing whether the moon was full at creation.[5] One of the arguments that others had made was that God must have created the moon fully formed. However, Augustine points out that the moon is always full, just not fully illuminated. As such, the moon was fully formed, whatever phase it was in. Crucially for our purposes, he argues that this is true whether the moon is self-illuminated or illuminated by the sun.

In his Exposition on Psalm 11, Augustine comments on the moon as an allegory of the Church. Psalm 11 does not mention the moon in the Hebrew, but does in the translation that Augustine was using. He explains two "probable opinions" about the light of the moon. The first is that the moon has its own light, but only half of the moon's sphere is lit. The changing phases of the moon result from the moon's revolving. The second is that the moon lacks its own light but is given light by the sun.

Augustine gives two different explanations of the allegory of the Church as the moon, corresponding to the two different theories about the moon's light. If the moon has its own light, it represents the split nature of the Church, combining the spiritual light of Christ corresponding to its light half with the carnal flesh of man corresponding to its dark half. On the other hand, if the moon's light comes from the sun, the allegory is that the light of the Church comes from the Son of God. Either way, Augustine sees the moon as an allegory for the Church.

---

[5] Augustine, *The Literal Meaning of Genesis. 1*, Book 2, Chapter 15.

John of Damascus

John of Damascus mentions the moon's lacking its own light.[6]
He insists that God could have granted it its own light, but he
created it this way to produce order and rhythm in nature and
teach us to live in community and under the authorities placed
over us.

Remigius of Auxerre

Remigius states in his commentary on Genesis[7] that "it has
been established that the moon is illuminated by the light of
the sun. itself." He also says this is part of the reason the sun
is greater than the moon.

Martin Luther

Luther discusses this subject in his commentary on Genesis.[8]
He indicates that astronomers say that both the stars and the
moon are lit by the sun. He indicates that this is proved by
a lunar eclipse where the earth's shadow falls on the moon.
Luther says that he does not either "reject or deny" this. However, he says that it was divine power that was able to create a
sun able to bring light to distant objects, as well as the moon
and stars able to receive this light.

Luther references Augustine's discussion of the moon as
an allegory for the Church. He describes Augustine as "forc-
ing" an allegory upon the Church while himself "defining
nothing." He states that we can learn from astronomers "what
points are possible to be disputed" within that science.

---

[6] John of Damascus, "An Exposition of the Orthodox Faith," Book 2, Chapter 15.
[7] Remigius of Auxerre, "Exposition on Genesis."
[8] Luther, *Luther On The Creation: A Critical and Devotional Commentry on Genesis*.

John Calvin

Calvin touched on this subject in his commentary on Genesis.[9] He insists that Moses knew full well that the sun lit up the moon. However, he instead stated that, as we perceive, the moon is a dispenser of light to us. Nevertheless, Calvin disputes the idea that the moon is entirely dark. Instead, he argues that it gives off light, but not enough to reach us. The sun merely supplies what the moon lacks. Calvin attempts to support his claim by referring to the science of his day. According to the classical conception, the universe was a series of spheres. The first sphere was earth, then water, then air, then fire, and possibly aether. The moon was located in or above the fiery sphere and, thus, by nature, must in fact be luminous.

Conclusions

Historical Christians

Historical Christians showed an awareness of the moon's not having its own light from very early on. Augustine alone seems to view it as a matter of scientific debate. However, there is no indication of anyone's seeing a conflict between this theory and the Biblical text until the reformers. Even Luther and Calvin do not explicitly identify this as an objection raised against the Biblical text but seem to implicitly acknowledge it as some sort of issue they need to discuss.

Calvin is the only author to give something like a response to this issue. He says that Moses spoke according to our perceptions, indicating that the moon dispenses light to us. However, he also thinks that the moon does give off light, just not enough to reach us by itself.

---

[9] Calvin, *Calvin's Commentary on the Bible.*

Modern Science

The idea that the moon did not have its own light was well established by the advent of modern science. It also became further clear that the planets in our solar system also reflect the light of the sun rather than emitting their own light. The stars, on the other hand, produce their own light. However, there is a complication: all objects emit light in what is called blackbody radiation. Objects such as the moon do technically emit a small amount of invisible light.

Evaluation of the Response

Historically, Christians have, for the most part, not seen a problem here and thus not addressed the issue. It never occurred to most of these historical authors that there was an issue. It is only at the time of the Reformation that we see some authors addressing this issue. The idea that the Biblical description of the moon as a light would be at odds with the moon's not generating its own light appears to be a relatively modern idea.

This issue illustrates the importance of going back to the source. We have to take the time to understand what the author meant by light. If we rush into the assumption that a light must be the original source of light, we run into problems. However, if we realize that in common parlance we use the word light to mean sources of light other than the original source, then there is no difficulty. We need to take care to understand what the Biblical text actually intended by its language.

Chapter 19

Are the Heavenly Bodies Alive?

The Scientific Question

The sun, moon, planets, and stars are heavenly bodies that move in the sky above. The whole sky, including all these bodies, revolves around the earth, but the sun, moon, and planets have their own additional movement. This movement is not the simple movement of falling to the ground, but a circular movement. Furthermore, even that circular movement is not constant. The planets will, at times, reverse direction and move in the opposite direction of their general movement. This complexity of movement is why they were termed wanderers or *planétēs* by the Greeks. But what was behind that movement?

A possible explanation of this movement is that the heavenly bodies are alive. They are some sort of living creature, which explains why they wander around the sky rather than follow the general movement of the heavens. This idea seems very strange to the modern mind, which is used to thinking of the planets as exhibiting the simple behavior of orbiting the sun. However, we must remember that if you simply watch the heavenly bodies from the perspective of the earth, the movement seems much more complicated.

There was a strong association between the gods and the heavenly bodies in the ancient world. Indeed, we still call the planets by the names of Roman gods. However, the exact relationship between the gods and the heavenly bodies in the minds of these ancient people is somewhat unclear. Moderns tend to assume that ancients believed that the heavenly bodies were the gods. This was probably true in some cases, but in other cases, the heavenly bodies were conceived of as representations of gods, possibly controlled by the gods but distinct from the gods themselves.

Greek philosophy typically moved away from polytheistic explanations of phenomena towards more naturalistic explanations. For example, Aristotle argued that the wind was caused not by wind gods but by air exhaled from the earth. Nevertheless, when it came to the heavenly bodies, the philosophers still invoked divinity. Plato thought the stars were some sort of divine beings. Aristotle thought instead that divine movers were responsible for turning the spheres, which caused the movement of the stars in the heavens.

The Biblical Discussion

Christians believe in only one God. However, we do not only believe in a single supernatural being. Rather, we believe in the existence, at least, of angels and demons. In fact, the word demon derives from the Greek word δαίμων (daimōn) which referred to a class of lesser deities. Consequently, a Christian might believe that the heavenly bodies, while not divine, were either angelic or moved by angelic creatures.

The Bible contains many passages that anthropomorphize the heavenly bodies, speaking of them as if they were living, rational creatures. For example, the psalmist calls upon the heavenly bodies to praise God.

> Praise him, sun and moon, praise him, all you shining stars! (Psalm 148:3 ESV)

One could take such a passage as indicating that these heavenly bodies are some sort of angelic or spiritual beings that can praise God. However, this Psalm calls upon all of creation, including fire, snow, and the hills, to praise God. Are all of these things spiritual beings? Presumably not.

Stars also make a frequent appearance in prophetic passages. A famous example is a passage from Isaiah that refers to a star fallen from heaven.

> How you are fallen from heaven, O Day Star, son of Dawn! How you are cut down to the ground, you who laid the nations low! You said in your heart, 'I will ascend to heaven; above the stars of God I will set my throne on high; I will sit on the mount of assembly in the far reaches of the north;' (Isaiah 14:12–13 ESV)

This passage is the source of the name Lucifer, often understood as the name of the devil. The term "day star" refers to Venus, which sometimes rises shortly before dawn. The Latin name for the morning star is Lucifer. That word appeared in the Vulgate and King James translations. The day star in this passage is often interpreted as the devil falling to earth. Consequently, the name Lucifer became associated with the devil.

Like most passages engaged in prophetic imagery, the meaning of these verses is disputed. As noted, some argue that this passage is describing the fall of Satan. However, these verses appear in a passage that describes itself as a "taunt against the king of Babylon," leading many commentators to argue that the passage is speaking only of a human king.

Whatever the correct interpretation, a star is being used as a symbol for either a mighty king or the devil.

Various other passages either engage in the anthropomorphization of heavenly bodies or use the heavenly bodies as part of their symbolic imagery. Neither usage gives us reason to think that the stars are alive. The Bible does make it clear that the sun, moon, and stars are not to be worshipped. Instead, they are things that have been given to everyone on earth:

> And beware lest you raise your eyes to heaven, and when you see the sun and the moon and the stars, all the host of heaven, you be drawn away and bow down to them and serve them, things that the LORD your God has allotted to all the peoples under the whole heaven. (Deuteronomy 4:19 ESV)

The Bible does not say much about the nature of the heavenly bodies besides describing them as lights. It makes it clear that they are not divine beings to be worshipped. There are poetic passages that use anthropomorphization and imagery; however, that does not give any reason to conclude that the heavenly bodies are anything but simple lights. Certainly, there is no Biblical indication that the heavenly bodies are alive.

Historical Christians

Origen

Origen stated that the Church has no clear teaching on whether or not the sun, moon, and stars are living beings.[1] However, he makes a scriptural argument that they are alive later in his book.[2] In connection with the creation of the heavenly bodies, he asks whether the sun, moon, and stars are rulers in the sense

---

[1] Origen, Cavadini, and Lubac, *On First Principles*, Preface.
[2] Origen, Cavadini, and Lubac, Book 1, Chapter 7.

of merely illuminating day or night or whether they rule as some sort of rational being.

He cites the book of Job:

> Behold, even the moon is not bright, and the stars are not pure (clean) in his eyes; how much less man, who is a maggot, and the son of man, who is a worm! (Job 25:5–6 ESV)

From this, Origen concludes that stars are subject to sin. In particular, the reference to the stars' being unclean must be a reference to their being infected with the contagion of sin. He argues that this cannot be a reference to simple cleanliness because that would mean that God had created dirty stars, which would reflect a defect in God's creative work. How can the text censure the stars for becoming dirty if they cannot, by their own diligent efforts, make themselves clean?

Origen then cites Isaiah:

> I made the earth and created man on it; it was my hands that stretched out the heavens, and I commanded all their host. (Isaiah 45:12 ESV)

Origen points out that God commanded the host of heaven. This indicates that God gave commands to the stars, which can only make sense if they are rational creatures able to receive and carry out commands. What commands? The commands to follow their observed orbits and movements.

Origen argues that the movement requires a soul. Furthermore, because the stars carry out their movement with such regularity, it is clear that an irrational being could not carry out such movements. Furthermore, it is only to be expected that a rational being would move both forward and backward, explaining the retrograde motion of planets where they sometimes move opposite to their regular orbits from the perspective of earth.

Origen cites Jeremiah as calling the moon, the Queen of Heaven. Jeremiah contains two passages which condemn the worship of the Queen of Heaven. But there is no indication there that this queen is the moon, or that the moon is a queen of any sort.

Lactantius

Lactantius argues that the stars are not gods.[3] He points out that they do not deviate from their prescribed orbits. A god would move "hither and thither in all directions" not in a particular path. Living creatures move wherever they want because wills are unconstrained and their movement voluntary. But the movements of the heavenly bodies obey laws appointed for them and are thus by necessity the voluntary movements of God.

Lactantius interacts with pagans who had argued that because the movement of the stars was not random it was therefore the voluntary movement of rational creatures. However, he insists that there is a third option besides random and voluntary movement, namely, movement by necessity. God, Lactantius argues, framed the world in such a way as to cause the stars to follow their appointed courses.

He references Archimedes of Syracuse whom he states was "able to contrive a likeness and representation of the universe in hollow brass, in which he so arranged the sun and moon, that they effected, as it were every day, motions unequal and resembling the revolutions of the heavens, and that sphere, while it revolved, exhibited not only the approaches and withdrawings of the sun, or the increase and waning of the moon, but also the unequal courses of the stars, whether fixed or wandering."

---

[3] Lactantius, "The Divine Institutes," Book II, Chapter V.

He is describing a mechanical model of the universe. The ancient Greeks had constructed machines that consisted of a hand crank connected to a complex system of gears and dials that represented the movements of the heavens as they understood them. One of these machines was discovered in a shipwreck in 1901 and is now called the Antikythera mechanism.

Lactantius's argument is that if humans can build this mechanical model, surely God could build a mechanical basis for the movement of the heavens. Nobody would insist that the figures of the stars in the mechanical model are intelligent beings or move themselves. Rather, they move by the genius of the artifice. Likewise, the stars are not alive or gods but move through the mechanical system devised by God.

Basil of Caesarea

Basil assumes without much comment that stars are not living beings while discussing astrology.[4] He objects to the idea that there are evil stars. If the stars are evil by their nature, then God is the author of evil. But this cannot be the case. In order for the stars to have chosen evil instead of being made evil, they would have to be animals of some kind. But Basil describes this as "the height of folly" and describes attributing the power of choice to the stars as "lies about beings without souls."

Augustine of Hippo

Augustine describes the question of whether the heavenly bodies are living creatures as one that was often asked.[5] He asks whether they are simply bodies without spirits, or are living creatures consisting of bodies united with spirits, or

---

[4] Basil of Caesarea, "Hexaemeron," Homily 6.

[5] Augustine, *The Literal Meaning of Genesis. 1*, Book 2, Chapter 18.

have spirits but not ones united to the heavenly bodies. He states that "This problem is not easy to solve." He urges caution in rushing to a conclusion on what the Scriptures say on this question. Otherwise, when new evidence arises, we may unduly reject the true theory by being too attached to an incorrect reading of the Bible.

Jerome

Jerome addresses this issue in his commentary on Ecclesiastes 1:5–6.[6]

> The sun rises, and the sun goes down, and hastens to the place where it rises. The wind blows to the south and goes around to the north; around and around goes the wind, and on its circuits the wind returns. (Ecclesiastes 1:5–6 ESV)

The same Hebrew word, *ruach*, means both wind and spirit. Jerome takes the second verse to refer to a spirit rather than the wind. He suggests that he may have referred to the sun as a spirit "on the ground that the sun is a living thing with breath and vigor, completing its annual cycle by motion of its own." Alternatively, he suggests that the author is referring to a spirit pervading all things. Jerome thinks that the sun is a living creature, or at least considers it a plausible possibility.

Cosmas Indicopleustes

Cosmas rejects the idea that the heavenly bodies are moved by the motion of the heavens, insisting instead that rational powers are responsible for moving them.[7] To support this, he quotes Paul:

---

[6] Jerome, *Commentary on Ecclesiastes*.
[7] Indicopleustes, *Christian Topography*, Book II.

in which you once walked, following the course of this world, following the prince of the power of the air, the spirit that is now at work in the sons of disobedience (Ephesians 2:2 ESV)

Cosmas takes this as indicating that before his fall, the devil's job was to move air around, hence being the prince of the power of the air. Other angels were tasked with moving the sun, moon, and stars. Still other angels prepared the clouds and the rain and performed many other services. All of these different tasks were given to angels in order to minister to the image of God—man.

John of Damascus

John wrote, "It must not be supposed that the heavens or the luminaries are endowed with life. For they are inanimate and insensible.[8] He explains Psalm 148's injunction for the heavens to rejoice as a command to the angels in heaven to rejoice. John points out that Scripture often uses personification, pointing to a specific passage from the Psalms:

The sea looked and fled; Jordan turned back. The mountains skipped like rams, the hills like lambs. What ails you, O sea, that you flee? O Jordan, that you turn back? (Psalm 114:3–5 ESV)

He mentions a phrase of his day, "the city gathered together," which means not that the buildings of the city gathered together but that the inhabitants gathered together. He also references Psalm 19:

To the choirmaster. A Psalm of David. The heavens declare the glory of God, and the sky above proclaims his handiwork. (Psalm 19:1 ESV)

---

[8] John of Damascus, "An Exposition of the Orthodox Faith," Book II, Chapter 6.

The passage does not mean that the heavens speak with an audible voice, but that their greatness brings to mind the power of the Creator.

Thomas Aquinas

Aquinas discusses whether the lights of heaven are living beings.[9] He notes disagreement among both philosophers and theologians. He mentions the story of Anaxagoras who is reported to have been condemned for teaching that the sun was a fiery mass of stone rather than a god. However, the Platonists believed that the heavenly bodies had life. He identifies Origen and Jerome as believing the stars had life. He notes that John of Damascus and Basil both believed that the stars were not alive, while Augustine left things uncertain.

In Aquinas's view, in order for the heavenly bodies to be alive, they would have to consist of a body and a soul. However, he argues that no purpose is served by uniting a soul to the heavenly bodies. Because the heavens are immutable in his view, there is no need or possibility of those bodies' receiving nutrition or growth which would require what Aquinas calls a nutritive soul. The heavenly body also cannot have senses according to Aquinas's view because in the classical science conception all senses worked on a principle of like sensing like. Because the heavens were made of aether, they could only sense aether. Intelligence, according to Aquinas, resides entirely in the soul and thus does not provide a reason for a body. Furthermore, the movement in the heavenly bodies can be affected by external forces and thus do not need souls to be placed in them.

---

[9] Aquinas, *The Summa Theologiæ of St. Thomas Aquinas*, The First Part, Summa 70, Article 3.

Aquinas argues that the heavenly bodies must move by the "direct influence and contact of some spiritual substance" rather than by nature. In the Aristotelian system, movement by nature was always movement to a proper place. Earth could fall because its proper place was at the center of the universe. However, circular movement such as those found in the heavens did not go to some proper place and then stop. Since this is not compatible with movement by nature, it must then be caused by some intelligent agent.

Aquinas goes on to argue that there is no actual disagreement between previous authors about whether or not the stars are alive. Instead, he argues that the distinction "is not a difference of things but of words." He does this by claiming that when the Platonists indicated that the heavenly bodies were alive, they only meant that they were moved by these spiritual substances.

John Buridan

Buridan mentions this issue while discussing what causes a thrown projectile to keep moving.[10] Aristotle had argued that the air pushed on the projectile, keeping it moving. However, Buridan rejected this idea, instead arguing that the projectile had an impetus imparted to it when it was thrown, which kept it in motion. This is a precursor to Newton's first law, which states that an object will remain in motion unless an outside force acts upon it.

Buridan notes that the Bible does not say that any sort of intelligence is responsible for moving the celestial bodies. As such, he suggests that God imparted impetus to the heavenly bodies thus keeping them moving without any further direct

---

[10] Buridan and Clagett, "Questions on the Eight Books of the Physics of Aristotle," Book VIII, Question 12.

intervention. However, Buridan does note that God remains active as a "co-agent in all things that take place." He references God's rest on the seventh day, indicating that this "committed to others" the function of keeping the universe going. By putting the impetus into the heavenly bodies, he could rest on the seventh day while the heavens remained in motion. He states that the impetus of the heavenly bodies was not decreased or corrupted because the heavenly bodies have no inclination for other movements, nor anything that resists that movement. However, he says that he does not give this opinion assertively, but rather tentatively, allowing the "theological masters" to correct him if necessary.

Dante Alighieri

Dante mentions the angels as intelligences which move the heavenly spheres.[11] The book in question consists of a poem and a commentary on that poem. Book 2's poem begins with the line:

You whose intellect the third sphere moves,

He explains in his commentary that he is referring to "certain intelligences" or angels who "preside over" the movement of the sphere of Venus. The moon is in the first sphere, Mercury in the second sphere, and Venus in the third sphere. Dante takes it for granted that the spheres are moved by angelic beings. He points out that the pagans took these heavenly bodies to be gods instead. The bulk of his discussion is concerned with the question of how many angels there are, arguing that there are numerous angels and not just a handful—one for each sphere.

---

[11] Dante Alighieri, *The Banquet*, Book 2.

Martin Luther

Martin Luther speaks disparagingly of the idea that the spheres were intellects or had an intelligent nature.[12] He refers to the influential Islamic philosopher, Averroes, describing his thoughts on the subject as "absurd." Averroes had attributed intelligence to the spheres because they moved in a perfect and regular manner. Luther attributes the "greatest and worst ignorance of God" to this idea. Luther does not elaborate on why he finds fault with this idea, but seems to view it as obviously absurd.

Conclusions

Historical Christians

The only Christian considered who thought that the stars were living beings was Origen. Lactantius, on the other hand, shows a surprisingly modern take in comparing the movement of the heavens to a mechanical machine. Basil, John of Damascus, and Jerome assume that the stars are not living creatures. Starting with Cosmas, we see the idea that while the stars are not alive, they are moved by the action of angelic beings. This idea is adhered to by Aquinas and Dante. Buridan suggests something akin to the modern understanding in proposing that the motion of heavenly bodies is due to the impetus present in these bodies. Luther scorned the idea of an intelligence behind the motion of the heavenly spheres.

Origen invokes Biblical passages that address the heavenly beings as if they were living creatures. John of Damascus points to other passages that engage in anthropomorphism to show that the passages do not mean that the heavenly bodies

---

[12] Luther, *Luther On The Creation: A Critical and Devotional Commentry on Genesis,* 1:Chapter I, Part II.

are alive. Other than this, we find little interaction with the Bible on this question. The nature of the heavenly bodies is either assumed or argued based on the science of the day of the author.

Modern Science

At the dawn of modern science, the heavens were conceived as a series of concentric spheres that something must be responsible for turning. However, this conception would be supplanted by the work of Kepler who reduced the movement of the planets to three laws of planetary motion. Isaac Newton would later show that these laws of motion were a consequence of his inverse-square law of gravitation and first law of motion (inertia). Consequently, the motion of the heavenly bodies was no longer a special motion where it made sense to invoke angelic beings.

It would still be possible to invoke angelic beings as the cause of gravity and inertia and, thus indirectly, of the heavenly bodies. It is conceivable that angelic beings are behind what we see as the ordinary laws of nature. However, such an idea is entirely speculative. Moreover, this would be crucially different from what the historical Christians thought about the heavenly bodies. According to them, the heavenly bodies were special and different and required the direct action of an intelligent agent to keep them in motion. That is an entirely different idea from the suggestion that the laws of nature are caused by spiritual beings in a general sense.

Evaluation of the Response

Origen is the sole author considered who believed that the heavenly bodies were actually living creatures. This was definitely incorrect; the stars are not living creatures. He does present a Scriptural case for his position. He attempts to argue from the statement that the stars aren't clean to their being alive, or else he takes a reference to a pagan deity, the Queen of Heaven, as an indication that the moon is alive. But these arguments are not compelling. Origen is early enough that he is primarily influenced by Plato, who thought of the heavenly bodies as divine beings. It seems likely that his conclusion is less based on the text and more due to the fact that he already believes the stars to be alive.

Several of the later authors were more influenced by Aristotle's ideas, and they thought angelic beings were responsible for the movement of the heavenly bodies without the heavenly bodies' being alive themselves. This would turn out to be incorrect. The only author to offer a Scriptural argument in favor of this idea is Cosmas, who attempts to get it from Paul's obscure reference to the "prince of the power of the air." For the most part, the authors either assume that this is the case or offer arguments in line with the science of their day.

Two authors, Cosmas and Origen, attempted to offer Scriptural arguments for their idea that stars were either living or directed by angelic creatures. The scriptural arguments are not very compelling. Nevertheless, they show that these authors considered the Bible to hold authority on these scientific questions. Origen and Cosmas both thought they could draw conclusions about the nature of the heavenly beings by appealing to Scripture.

Others rejected the prevailing ideas of their day. Basil appears to take it for granted that the heavens are not living creatures. John of Damascus rejects the idea that they are alive. Neither suggested that the heavens were moved by angelic beings. Lactantius shows a surprisingly modern take in comparing the heavens to a mechanical device. He does this as part of rejecting the pagan understanding of the heavenly bodies as gods. Buridan anticipates Newtonian mechanics as an explanation for the movement of the heavenly bodies. He connects the idea of God's placing impetus in the heavenly bodies to keep them going to God's resting from his acts of creation.

All of these authors went back to the source. They thought about the world from a Christian perspective. Instead of simply reworking pagan ideas about the movements of the heavens, they understood the heavenly bodies as non-living lights created by God.

Chapter 20

Do the Stars Determine Our Actions?

The Scientific Question

It is abundantly clear that at least some heavenly bodies have significant effects on the earth. The sun is the most obvious, giving warmth and light to the earth. It is the primary cause of the growth of crops. The moon also induces certain effects. In addition to the tides, oysters, for example, open and close in sync with the phases of the moon. It would not be strange to think that the other visible heavenly bodies have similar effects on the earth.

Ancient people observed that certain events could be reliably predicted based on the appearance of certain stars. When the ancient Egyptians saw the star Sirius rise in the east, they knew that the Nile river was about to flood. When the ancient Greeks saw the Pleiades, a cluster of seven stars, rise, they knew the sailing season had begun. The Romans noticed that the hottest weather happened when the sun was located in the constellation of Canis Major, the Great Dog. This time of year would eventually become known as the dog days of summer.

Furthermore, the heavens often showed signs that were much more dramatic than the simple rising and setting of stars.

Sometimes, the sun will go dark in the middle of the day. The moon will turn red. Stars moving in one direction will slow, stop, and move in the opposite direction. New stars will briefly light up and fly across the sky. Comets, in particular, were seen as evil omens because, unlike the rest of the movements of the heavens, they appeared erratic and unpredictable. If the relatively ordinary movements of the stars had effects on the earth, how much more effect would these dramatic signs have?

This idea would lead to what we would today know as astrology, the attempt to predict future events based on the stars. A system of astrology arose in Greece that was a synthesis of astrological ideas from Egypt and Babylon. It places a particular significance on the positions of the heavenly bodies at the exact time of one's birth. This is called a person's horoscope or nativity.

There were stories that seemed to confirm that, at least in some cases, astrological predictions were correct. For example, Alexander the Great was warned by Persian astrologers that if he entered Babylon, he would die there, which he did. Pericles launched an expedition in the Peloponnesian War despite an eclipse, and he died a couple of years later and his city, Athens, went down to defeat.

The definitive ancient text on astrology was Ptolemy's *Tetrabiblos* or "Four books." Ptolemy is also known for his work, the *Almagest*, which set forward the classical understanding of the motion of the stars and planets. For him, the study of the movements of the heavenly bodies, astronomy, and the study of the effects these movements had on earth, astrology, were two complementary sciences.

Nevertheless, astrology was not without its ancient critics. Most notably, a school of philosophy known as the Academic Skeptics was critical of astrological prediction. They emphasized the human inability to know the truth about anything with certainty. They argued that a belief in fate or causal determinism undermined human morality and free will. They pointed out that twins born at the same time have different fates, and people born at different times will nonetheless sometimes end up dying in the same battle.

Today, the word astrology refers specifically to attempts to predict human actions and events on the basis of the stars. It is contrasted with astronomy, the simple study of the movements of the heavenly bodies. The distinction between these two existed at least as long ago as Ptolemy; however, the terminology is relatively modern. Historical authors would use other terminology to distinguish between these two fields. For simplicity, this chapter will use modern terminology throughout.

The Biblical Discussion

The Bible condemns divination, fortune-telling, and the interpretation of omens:

> There shall not be found among you anyone who burns his son or his daughter as an offering, anyone who practices divination or tells fortunes or interprets omens, or a sorcerer... (Deut 18:10 ESV)

Jeremiah warns his readers not to be dismayed at the signs of the heavens:

> Thus says the LORD: "Learn not the way of the nations, nor be dismayed at the signs of the heavens because the nations are dismayed at them" (Jeremiah 10:2 ESV)

Isaiah warns of coming doom and disaster, indicating that it will come unexpectedly; he tauntingly suggests his audience should look to salvation from the astrologers who were unable to predict the coming doom.

> But evil shall come upon you, which you will not know how to charm away; disaster shall fall upon you, for which you will not be able to atone; and ruin shall come upon you suddenly, of which you know nothing. Stand fast in your enchantments and your many sorceries, with which you have labored from your youth; perhaps you may be able to succeed; perhaps you may inspire terror. You are wearied with your many counsels; let them stand forth and save you, those who divide the heavens, who gaze at the stars, who at the new moons make known what shall come upon you. (Isaiah 47:11–13 ESV)

On the other hand, the Genesis 1 creation account states the heavenly bodies were created "for signs":

> And God said, "Let there be lights in the expanse of the heavens to separate the day from the night. And let them be for signs and for seasons, and for days and years" (Genesis 1:14 ESV)

This has sometimes been argued to support astrology by indicating that the heavenly bodies were created to be signs of future events. However, the clear rejection of such interpretations in other passages indicates that this is not what the passage has in mind. Most likely, the passage indicates that they are signs marking seasons, days, and years.

The Biblical perspective seems pretty clear. It condemns attempting to predict the future. It says not to follow the fears of the nations based on what they see in the heavens. It mocks those who attempted to predict the future based on signs in the heavens.

Historical Christians

Origen of Alexandria

We have extracts of Origen discussing astrology[1] in the context of explaining the meaning of the heavenly bodies as signs. He notes that many outside of the Church incorrectly think that everything that happens is caused by the position of the planets in the constellations of the zodiac. He suggests that even many Christians are distracted by the thought that human affairs might be controlled by the stars.

Origen insists that the stars are not the cause of human actions. His primary objection is that this idea undermines free will and personal responsibility. If one's actions are the result of stars, nobody is responsible for their own actions. Even whether or not someone believes in fate or accepts the Christian faith is a consequence of the positions of the stars. Consequently, there can be no justice in any human or divine punishment.

Origen points out that the astrologers not only sought to predict future events but also claimed the ability to ascertain information about the mother, father, and older siblings of a person. But those events took place before the person was born and because cause must precede effect, the stars at the time of one's birth cannot have caused those events. Furthermore, many events involve numerous people. Which set of stars was the cause of those events?

He further compares astrology to augury, the practice of interpreting omens related to birds, which was not thought to cause any events but merely to foretell them. It was clear that, at least some of the time, the stars are at most foretelling events and not causing them. He insists that it is silly to suggest

---

[1] Origen, *The Philocalia of Origen*, Chapter XXIII.

that the stars sometimes cause events while other times only indicating or foretelling them.

However, Origen does not object to the idea that the stars may predict future events. Since God has foreknowledge, he can send signs through the stars predicting or prophesying future events without those stars' being the cause of the events. Origen's objection is to the stars' controlling our actions, not to their predicting our actions.

While Origen rejects the idea that the stars cause our actions, he accepts that they are signs of future events. However, they are not signs to humans but to "powers." He is likely referring to texts such as:

> For we do not wrestle against flesh and blood, but against the rulers, against the authorities, against the cosmic powers over this present darkness, against the spiritual forces of evil in the heavenly places. (Ephesians 6:12 ESV)

The stars are signs for spiritual powers, giving them information and instructions.

Origen cites a now-lost work known as the *Prayer of Joseph*. We do not know much about this book, aside from a few quotations and references. The largest surviving fragment is an elaboration of the story of Jacob's divine wrestling match. However, in this work, Jacob is himself an angel and is wrestling the angel Uriel.

Here, Origen quotes the *Prayer* as saying:

> For I read in the pages of the sky what shall befall thee and thy sons.

The text indicates that Jacob was able to determine future events from his observations of the stars. Origen also quotes, "The heavens shall be rolled together as a book," which could be a reference to a couple of different verses:

> All the host of heaven shall rot away, and the skies roll up like a scroll. All their host shall fall, as leaves fall from the vine, like leaves falling from the fig tree. (Isaiah 34:4 ESV)

> The sky vanished like a scroll that is being rolled up, and every mountain and island was removed from its place. (Revelation 6:14 ESV)

These verses are typically taken as describing the destruction of the heavens. However, Origen takes it as suggesting that the heavens are like a scroll or book that conveys information. He also cites Isaiah's warning not to be dismayed by the signs of heaven, apparently taking it as implying that the signs convey accurate information.

However, while greater spiritual powers are able to understand the signs of the heavens, Origen argues that humans are incapable of understanding them. He argues that in order to have an accurate astrological forecast, we would have to know very precisely the time of birth. In an age long before mechanical clocks, such precision would be impossible.

He also points out axial precession, a phenomenon where the earth's axis slowly shifts in relation to the fixed stars. The consequence is that, over time, stars appear earlier and earlier in the year. However, the systems of astrological predictions were made based on the times of the year in which the stars appeared when astrology was first developed. This was no longer completely accurate in Origen's time but is much more out of sync now.

He points out that, in many cases, there are multiple signs involved that are claimed to blend together. But the interaction of so many different signs makes the whole enterprise of astrology very complicated. Origen says that those who study astrology "must despair of understanding such matters."

Origen argues that these difficulties make it impossible for humans to ascertain the meaning of the signs of the heavens. Indeed, Origen states that those with experience of astrological predictions know that they fail more frequently than they succeed. Origen quotes Isaiah:

> You are wearied with your many counsels; let them stand forth and save you, those who divide the heavens, who gaze at the stars, who at the new moons make known what shall come upon you. (Isaiah 47:13 ESV)

He takes this verse as indicating the inability of man to understand the signs and make accurate predictions.

Basil of Caesarea

Basil discusses the signs found in the heavenly bodies.[2] He mentions signs related to weather prediction. As an example, he gives the words of Jesus:

> He answered them, "When it is evening, you say, 'It will be fair weather, for the sky is red.' And in the morning, 'It will be stormy today, for the sky is red and threatening.' You know how to interpret the appearance of the sky, but you cannot interpret the signs of the times." (Matthew 16:2–3 ESV)

This is the same idea that is expressed in the adage:

> Red sky at night, sailor's delight. Red sky in morning, sailor's warning.

He also mentions other signs such as halos around the sun or moon, false suns, or rainbow colors on clouds. Seeing these signs is an indication of coming storms. He also mentions subtly different appearances of the phases of the moon, indicating different sorts of weather. Basil points out how useful these signs are for humans. He also mentions that at the dissolution

---

[2] Basil of Caesarea, "Hexaemeron," Homily 6.

of the universe, there will be signs in the heavens, namely the sun being turned to blood and the moon not giving her light. Basil's citation is actually reversed from the Biblical reference; it is the moon that turns to blood.

However, Basil indicates that those who would cite the Scripture's description of the stars as a sign in defense of their practice of astrology "overstep the borders." They see in the text not simple signs of weather or changes in the season but, "at the will of their imagination," human destinies.

Basil points out that in order to have an accurate astrological prediction, it would be necessary to have the precise time of birth. Even the difference in time between the birth and the time the nurse announces the birth could be sufficient to change the astrological prediction. Furthermore, it makes no sense to attribute the nature of the different animals of the zodiac to people born while the sun is in that portion of the sky. If stars dictate destinies, why are not kings born whenever the right configuration occurs? Why, in a dynasty of kings, does none of them happen to be born under stars condemning them to a life of slavery?

Basil argues that if our actions are caused by the stars, there can be no blame for those actions. It makes no sense to punish the thief or other criminal for actions that they have no control over. Even planting a field with seed is pointless, because whether or not the field grows is solely a consequence of the motion of the stars. For Christians, a world with fate removes our hope. If man does not act with freedom, there can be neither judgment nor just reward.

Augustine of Hippo

Augustine discusses this issue while considering the cause of the greatness of the Roman Empire.[3] He argues that it was neither a chance event nor fate. Rather, the Roman Empire was established by divine providence. One might refer to God's providence as fate, but Augustine insists that this would be incorrect terminology.

Augustine argues that if one believes that the stars determine the fates of men and empires without themselves depending in any way on a divine will, then one can be neither a Christian nor a worshipper of any other, false god. If the stars control everything, there can be no god who is worthy of praise and worship.

However, Augustine is more concerned about those who would maintain that while the stars control human actions and events, the stars themselves are under the control of God. But what of the evil done by men on earth? If evil is done because of the stars' acting on some will of their own, how can God judge men's actions? If the stars are not operating on their own will but on the direction of God, this makes God the author of all evil. It cannot be that the actions of men are controlled by the stars.

Augustine allows that it would be possible for the stars to signify or foretell events without causing them. However, this was not what the astrologers claimed; they thought the stars did not merely foretell events, but actually caused them.

Astrologers could not explain why twins, born at the same time, could nevertheless have very different lives. Indeed, the twins might even be of different sexes. Augustine specifically cites the very different lives of Jacob and Esau as examples. The astrologers would answer that the heavens changed so quickly that even the short interval between the births of both twins was sufficient to change his astrological prediction. However,

---

[3] Augustine, "The City of God," Book V.

if predictions change that much over a short period of time, how can any prediction be meaningful?

Augustine considers a historical report in which two twin brothers took ill, displayed the same progression of symptoms, and then recovered at the same time. The astrologers explained this as a consequence of both brothers' being twins who were conceived and born under the same stars. However, Augustine argues that it is much more plausible to think that, after having lived similar lives from birth to adulthood, they reacted in a similar way to the same disease.

Augustine also mentions the signs in his commentary on Genesis.[4] He insists that the signs do not refer to "what foolish men observe," but rather are practical and necessary signs like the signs that mariners use to steer their ship or which indicate changes in weather. Even our measurement of time itself derives from the movements of the heavenly bodies.

John of Damascus

Damascene discussed this issue while commenting on the creation of the heavenly bodies.[5] He insists that we get signs of rain, drought, cold, heat, moisture, and dryness from these bodies but not human actions. Rather, we have been created with free will. If, instead, our actions were caused by the stars, there could be neither virtue nor vice to our actions. We could thus deserve neither praise nor punishment. Even God must then be unjust for punishing some and rewarding others.

However, Damascene allows that the stars might be signs of events such as wars without being the cause of them. Furthermore, he suggests that the "quality of the air" produced by the heavenly bodies brings about various temperaments, habits,

---

[4] Augustine, *The Literal Meaning of Genesis. 1*, Book II, Chapter 4.
[5] John of Damascus, "An Exposition of the Orthodox Faith," Book II, Chapter VII.

and dispositions. However, human reason rules over, directs, and changes these characteristics.

According to Damascene, comets are the signs of the deaths of kings. According to to him, comets are not any of the ordinary stars but are formed by divine command and soon dissolve.

Thomas Aquinas

Aquinas discusses astrology in a couple of places.[6] Aquinas partially accepts the logic of classical astrology. He thinks that events on earth are caused by the heavenly bodies, with two classes of exceptions: random events and free will.

Aquinas argues that if our will and intellect were located in our physical bodies, we would have no free will but would simply act by instinct like an animal. In such a case, our actions would be governed by cause and effect and ultimately by the heavenly bodies. However, Aquinas takes it as manifestly clear that we have free will and thus our will and intellect are not located in our physical bodies.

However, this does not mean that human actions are unaffected by the heavenly bodies. Our wills and intellects are affected by the heavenly bodies through our senses. These effects may incline humans towards certain acts. However, the will can override natural human inclination.

In response to the argument that astrologers make accurate predictions, Aquinas seems to agree. He explains this as most men's failure to override their natural inclinations. A man can overrule his passions but in most cases, most men do not. As such, most men act the way astrologers would predict. He also suggests the action of demons may be involved in bringing

---

[6] Aquinas, *The Summa Theologiæ of St. Thomas Aquinas*, First Part, Question 115, Article 4 and The Second Part of The Second Part, Question 95, Article 5.

some predictions to pass. Aquinas rejects the idea that the stars could signify events without being their cause. He insists that a physical sign must have a causal relationship with the thing it signifies.

For Aquinas, there is nothing wrong with using the stars to make predictions about things that actually follow by necessity from the stars. For example, there is nothing unlawful about predicting an eclipse or weather by observation of the stars. However, it is unlawful to attempt to use astrology to predict the outcomes of random events. As for the actions of humans, Aquinas indicates that it is wrong to attempt to use astrology to gain "certitude" about their future actions.

Martin Luther

While commenting on Day Four of creation, Luther asks about the meaning of "And let them be for signs."[7] He notes that some had explained phrase as referring to signs' signifying weather, such as rain or storms. However, Luther expresses doubts that the heavenly bodies actually indicate such events with any certainty. He mentions Jesus' indication that a red sky in the morning or evening indicates foul or fair weather. Luther also mentions the idea that the Pleiades indicate rain and other similar proverbs. Luther indicates that he is not strongly opposed to these ideas, but he is also not willing to completely accept them because they are not uniformly sure or certain signs.

Luther suggests that the "simplest meaning of the text" is that it refers to signs such as eclipses or collisions of heavenly bodies. The signs are intended as wonders and miracles whereby God reveals his wrath. This includes meteors, falling

---

[7] Luther, *Luther On The Creation: A Critical and Devotional Commentry on Genesis*, 1:Vol 1, Section II, Part 4.

stars, halos surrounding the sun and moon, the appearance of rainbows, etc.

As to the signs' possibly being examples of astrology, Luther says:

> If they are not defended perseveringly and pertinaciously, I will not combat them strongly.

He indicates that he is not persuaded to consider astrology a science. He notes that they record their successful predictions but forget about their unsuccessful predictions. Indeed, their predictions fail very often. He insists that they cannot make a science out of a few cases where they got their predictions right.

Philip Melanchthon

Melanchthon wrote a defense of astrology in a preface to a book entitled, *On The Judgment of Nativities*.[8] He distinguishes between what we would today call astronomy and astrology. He argues that we know the effects of the heavenly bodies through continual experience, quoting Galen:

> It has been shown by manifest experience that there are effects of the stars and light on the elements and mixed bodies, and on plants and animals. As we know that fire is hot, water moist (for so I speak as an example), so we know that bodies are warmed by the Sun and moistened by the Moon, that the conjunction of Saturn and Mars makes hot spells and dryness, and that the conjunction of the Moon and Jupiter in Cancer increases moisture.

He insists that the practice of astrology is natural and not a superstition. There is no difference in essence between study-

---

[8] Melanchthon, "Preface."

ing the effects of herbs on the body and the effects of stars on the temperament.

Melanchthon thinks that the stars have an effect on a person's temperament and inclination but he does not believe that all of a man's actions are due to the stars. He points to three other causes he holds to be involved in the actions of man: human will, God, and the devil. He insists astrological divination is a good art precisely because understanding the inclinations of nature better enables us to control those inclinations.

Melanchthon declares that explaining Jeremiah's command not to fear the signs of heaven is easy. Jeremiah does not say that the signs are nothing, instead implicitly affirming them as signs. Rather, the command is not to be frightened of the events that these signs portend. We should rather be confident in God.

Melanchthon engages with Basil of Caesarea's objection that if the stars cause the sinful actions of man, this reflects poorly on God, their creator. In response, Melanchthon states that the stars had different effects on man before our nature was corrupted. It is only with our corrupted nature that the stars began to give us an inclination towards sin. Furthermore, the stars alone are not the cause of our actions.

He acknowledges that there are very few "demonstrations of the effects of the stars." He notes that astrologers are often deceived, even more so than practitioners of other arts. He attributes this to the weakness of the human mind. He nonetheless insists that astrology is not nothing and even though it cannot predict the future perfectly, it can predict some things, and even predicting some things is useful.

341

John Calvin

John Calvin discusses the meaning of the signs while commenting on the creation account.[9] He speaks disparagingly of "certain inquisitive persons" who "abuse this passage" to lend support for their "frivolous predictions." Calvin insists that these signs are about things belonging to the order of nature. In support of this view, he refers to Isaiah's and Jeremiah's condemnation of astrology.

In addition, Calvin wrote a treatise against astrology entitled *A Warning Against Judiciary Astrology and Other Prevalent Curiosities*.[10] By judiciary astrology, Calvin is referring to what we today would call astrology rather than astronomy, which he refers to as true astrology. He has harsh words for the practitioners of astrology, calling them "shameless deceivers" and astrology a "diabolical superstition."

Nevertheless, Calvin does not completely reject the influence of the stars. He thinks the stars have some influence over a person's nature, although this is not the primary influence. He sees merit in doctors' considering the positions of the heavenly bodies in determining the appropriate timing of treatments. He also suggests that God sometimes sends comets or other heavenly signs to warn of his coming judgments.

Calvin gives a number of arguments against astrology. He points out that many people born at the same time have very different lives. Even twins can be different in every way, including sex. In certain battles, numerous people die who were born at very different times. He points out the impracticality of knowing the state of the stars at the exact time of birth. Further, he argues that it would make much more sense to use the time of conception rather than the time of birth.

---

[9] Calvin, *Calvin's Commentary on the Bible*.

[10] Calvin, "A Warning against Judiciary Astrology and Other Prevalent Curiosities."

In support, he cites Biblical passages, including Isaiah's mockery of astrologers, Jeremiah's admonition not to be dismayed at the signs of the heavens, and Deuteronomy's prohibition of divination. He also cites Daniel:

> Daniel answered the king and said, "No wise men, enchanters, magicians, or astrologers can show to the king the mystery that the king has asked, but there is a God in heaven who reveals mysteries, and he has made known to King Nebuchadnezzar what will be in the latter days. Your dream and the visions of your head as you lay in bed are these: (Daniel 2:27–28 ESV)

Calvin goes to engage arguments put forward by the defenders of astrology. These astrologers pointed to Genesis 1, referring to the heavenly bodies as signs. Jeremiah also calls them signs when he warns his hearers not to be dismayed at them. From this, the defenders of astrology argued that the text does indeed indicate that heavenly bodies are signs, as astrology claims.

In response, Calvin points out that it is common to use the terminology of one's opponent. The Bible refers to the idols as gods, even though they are not gods. Likewise, Jeremiah can refer to the heavenly bodies as giving signs, even though they are not real signs. He also explains that they can be signs in the sense of guiding the right time to sow, plant, heal, give medicine, or cut trees.

The defenders of astrology also argued that Joseph, Moses, and Daniel were all educated in the astrological teachings of the Egyptians or Chaldeans. There are a couple of references in the Bible to Joseph's practicing divination:

> They had gone only a short distance from the city. Now Joseph said to his steward, "Up, follow after the men, and when you overtake them, say to them, 'Why have you repaid

evil for good? Is it not from this that my lord drinks, and by this that he practices divination? You have done evil in doing this.'" (Genesis 44:4–5 ESV)

Joseph said to them, "What deed is this that you have done? Do you not know that a man like me can indeed practice divination?" (Genesis 44:15 ESV)

However, this takes place during an episode where Joseph is keeping his identity concealed from his brother and is playing a ruse on them. As such, Calvin argues that Joseph's claims to practice divination are part of his ruse and not his actual practice. Furthermore, Joseph did not predict the famine by the stars, but by miraculous revelation. The Egyptians, who studied the stars closely, knew nothing of the coming famine.

Moses was claimed to practice astrology due to his description in Acts:

And Moses was instructed in all the wisdom of the Egyptians, and he was mighty in his words and deeds. (Acts 7:22 ESV)

Arguably, all of the wisdom would have included their practice of astrology. Calvin responds that this would not include their superstitions, and in any case, we have no indication that Moses used this astrology, even if he was trained in it. Rather, Moses always "governed himself by the word of God," doing what God commanded and not at any point attempting to lead Israel according to the stars.

Daniel was claimed to have practiced astrology on the basis of having undergone a three-year education in Babylon. However, Calvin points out that there is no indication that Daniel wanted to know anything about astrology as opposed to simple astronomy.

Astrology's defenders would also invoke Jesus' prediction of signs in the heavens announcing the days of his final coming. He compares these people to churchmen talking about military affairs. Just as a pastor would have no ability to competently discuss military tactics, these astrologers lack the competence to discuss the Scriptures.

Calvin declares that the events described by Jesus are extraordinary events that have nothing to do with the stars. They are not the ordinary natural movements of constellations. The prophets use similar language to express the coming anger of God. As regards the last day, Calvin declares that there will be visible signs, but whatever exactly happens, these signs are not the moving positions of ordinary stars upon which the astrologers "pretend to found their art."

Conclusions

Historical Christians

We have only looked at a sampling of the interaction between Christians and astrology. Indeed, a whole other book could probably be written investigating the intricacies of that interaction. However, we have looked at the most influential of the authors and thus should gain a general idea of the character of the interaction.

The pre-Reformation authors all argue against astrology because if human actions were controlled by the stars, we would not have free will. Origen, Basil, John of Damascus, and Augustine argue in particular that there can be no justice to blame or punishment against people who have no control over their actions. Origen and John of Damascus are willing to accept the idea that the stars predict human actions but insist they do not cause them. Calvin and Luther do not address free will. Melanchthon defended a version of astrology that was not fatalistic.

Some authors granted some degree of accuracy to astrology. Origen thought the stars were indeed signs, including some of future events, but these signs were not intended for humans. John of Damascus thinks that a person's temperament is affected by the heavenly bodies. Aquinas thinks that the stars do influence human actions sufficiently to make astrological predictions somewhat accurate. Calvin and John of Damascus both believe that God sends comets as signs. Calvin thinks that a person's nature is affected by the stars, even though it is not their primary influence. Melanchthon's position is not too different from that of Calvin, but he thinks the influence is stronger.

There was some difference of opinion on the sense in which the heavenly bodies were signs in the Genesis 1 creation account. Origen thinks they are God's method of communicating with spiritual powers. Basil, Augustine, and John of Damascus thought the text referred to signs of coming weather. Luther believes they are a reference to unusual astronomical events like the collisions of heavenly bodies or eclipses.

We see a number of recurring arguments against the accuracy of astrology. It is impractical to obtain the exact time of a person's birth. Twins often lead very different lives. People born at different times nevertheless end up dying in the same battles. It was noted that astrological predictions were more often wrong than they were right. Even Melanchthon, who defended astrology, acknowledged its limited accuracy.

Modern Science

The astronomers who played key roles in the rise of heliocentrism were all practicing astrologers. At the advent of modern science, astrology was widespread and even considered a respected part of science. However, modern science would end up undermining the logic behind astrology. Whereas in classical science, the day, night, and seasons were caused by the motions of the heavens and the heavenly bodies, in modern science they are caused by the rotation and orbit of the earth. It is the movement of the earth and not the heavens that is responsible for day, night, and the seasons.

In the context of the emergence of modern science, it became increasingly clear that astrology simply did not work. For example, Newton's theory could give precise predictions on the movements of physical objects. But astrology only occasionally managed to work. Furthermore, as the tenets of classical science were rejected, astrology transitioned from the insight of the lost great ancient civilization to just another incorrect idea of the ancient world.

Astrology would see a resurgence in popular culture during the twentieth century. As of 2018, 30% of Americans believed in astrology. Nevertheless, the scientific community considers astrology to be completely baseless.

Evaluation of the Response

Historical Christians rejected astrology in its fullest form. However, they varied in their acceptance of the accuracy of astrological predictions. Some rejected it completely, while others thought it was simply beyond human ability, had limited accuracy, or was usually accurate. They also differed in terms of whether they thought that human actions were caused in some limited way by the movement of the stars.

The conclusion of modern science is that astrology has no accuracy. Those who allowed that there might be some accuracy to astrological predictions or that the stars had some effect on a person's nature have turned out to be granting too much to the astrological ideas of their day. Even Calvin's vociferous attack on astrology does not go so far as to entirely reject it.

We do see respect for the authority of the Biblical text in these historical authors. In some cases, we see them rejecting astrological ideas on the basis of Biblical teaching. Others attempted to show that Biblical teaching was compatible with the practice of astrology. Even though we would reject their argument and find their reasoning dubious, this still shows that they saw the Biblical text as having authority.

Nevertheless, they ought to have gone farther in rejecting these ideas. The Bible was clear in its condemnation of astrology and all forms of fortune-telling. Nevertheless, many Christians defended it to one degree or another. It was also widely understood that astrologers made many inaccurate predictions. They failed to return to the source. They ought to have gone back to the Scriptures and seen that it completely rejected astrology and they ought to have gone back to the evidence and seen that there was no reason to find astrology credible.

Chapter 21

What Was the Star of Bethlehem?

The Biblical Question

Consider this telling of the story of the Star of Bethlehem: three sages see a new star that has never been seen before. It is brighter and more beautiful than any of the other stars. As the sages attempt to get closer to this sight in order to see it better, the star begins, to their great surprise, to move. The sages follow this moving star and it leads them from their country in the far east to Jerusalem. During the day, the star is bright enough that it can still be seen so they can follow it. When they arrive in Jerusalem, the star disappears. So they ask the people of Jerusalem, "Where is he who has been born king of the Jews?" Herod has his scholars look through their Scriptures and determine that the Christ would be born in Bethlehem. The sages set out for Bethlehem, and as they do so, the star reappears, and they follow the star to Bethlehem. Then the star descends on the very house where Jesus is staying, and the sages enter and give him gifts.

Consider another, quite different version. A group of astrologers observed a star's heliacal rise. The heliacal rise is when a star first becomes visible just above the eastern horizon (that is, near to the sun). For most people, the event was unimpressive and passed without comment. However, because of their practice of astrology, they concluded that this event was significant. They concluded that it indicated that a new king of the Jews had been born, and they set out to see him. They went to where one would expect to find the new king of the Jews, the capital of Judea, Jerusalem. The local scholars report that the powerful king they seek is to be found in Bethlehem. The reigning king, Herod, sends them to Bethlehem, telling them to seek out the king. So they set out towards Bethlehem. Once they reach Bethlehem, they observe the same sign in the heavens again. They ask around the city and eventually find the infant Christ and give him gifts.

Chances are the first sounds familiar, although perhaps containing a few elements you are uncertain of. Nevertheless, the second account sounds bizarre and is not the way you envision the Bethlehem Star at all. But what does the actual Biblical account say?

> Now after Jesus was born in Bethlehem of Judea in the days
> of Herod the king, behold, wise men from the east came to
> Jerusalem, (Matthew 2:1 ESV)

The visitors are described as being "wise men from the east." The Greek word translated as "wise men" is μάγοι (*magoi*), sometimes it is also transliterated as "magi." The term magi originally referred to priests of Zoroastrianism, a religion practiced in what is now Iran (formerly Persia), a region that was in the east. They were notable for their practice of astrology. The term came to refer to magicians, astrologers, and sorcerers. The only other use of the term in the New Testament is to refer to magicians in the book of Acts.

> saying, "Where is he who has been born king of the Jews? For we saw his star when it rose and have come to worship him." (Matthew 2:2 ESV)

The Magi report seeing his star when it rose. Some older translations say instead that they saw his star in the east. The Greek word is ἀνατολῇ (*anatolēi*) which means either rising or the direction in which the sun rises, the east. The Vulgate translated it as east, but modern scholars agree that it should be translated as rising. This plausibly refers to the heliacal rising of a star.

However, what is notable is the absence of certain aspects of the common story. There is no indication that this was a new star that had not been seen before. They also do not speak of the star's being especially bright or beautiful. Certainly, the text does not speak of the star's being visible during the day. Additionally, there is no indication that they followed the star to Jerusalem.

The idea that they were following a star to Jerusalem does not make sense. If they were following a star, they would not need to go to Jerusalem. They would have simply followed the star to Bethlehem. Furthermore, whether one takes the text to say that the they saw the star in the east or in its helical rising, the star would have been in the east, and would have led these Magi from the east to their east, not west towards Jerusalem.

> Then Herod summoned the wise men secretly and ascertained from them what time the star had appeared. And he sent them to Bethlehem, saying, "Go and search diligently for the child, and when you have found him, bring me word, that I too may come and worship him." (Matthew 2:7–8 ESV)

Herod does not merely tell the Magi where the Christ was to be born, but sends them to Bethlehem with instructions to search diligently for the child. Herod is not expecting miraculous assistance in finding the house where Jesus was staying. The word translated as "search" here is translated as "question" or "inquire" in other places. He is expecting them to ask appropriate questions to locate this child. Probably, he is sending the Magi because he knows that if he sent his soldiers, the people of Bethlehem would be unlikely to be co-operative with his investigation.

> After listening to the king, they went on their way. And behold, the star that they had seen when it rose went before them until it came to rest over the place where the child was. When they saw the star, they rejoiced exceedingly with great joy. (Matthew 2:9–10 ESV)

The star is described as going before them and then coming to rest over the place where the child was. This is often taken to mean the Magi followed this heavenly light from Jerusalem to Bethlehem and to the very house where Jesus was staying. The word translated as "went before" is προάγω (proagō). It does not generally mean to lead, but merely to go first. It actually does not make much sense that the Magi would need a star to guide them to Bethlehem, because they were already heading to Bethlehem after having been sent by Herod.

Furthermore, those in the ancient world would have thought of the stars as being in motion. Over the course of a night, the stars circle around the north pole, with some stars rising and others setting. From night to night, the stars move to slightly different positions. Planets, which would have been classified as a kind of star, move against the background of the other stars. As such, it is possible that the text may be referring to these ordinary movements, not to some extraordinary movement.

The text describes the star as coming to rest. Planets do, in fact, stop moving. When the earth passes an outer planet in its orbit, that planet appears to slow, stop, and then reverse direction. It is possible that such a cessation of movement is what the text is referring to. Alternatively, the text literally says that the star came and stood over the place where the child was. Given how slowly stars move, it would not be odd to describe them as standing. Under this interpretation, the text merely indicates that the star had made its way over that place without indicating that it had completely ceased moving.

The star comes to rest over the place where the child was. This has often been taken as indicating that the star was somehow directly over the house where Jesus was staying. However, the text does not indicate the scope of the place. It could be a house, or the village of Bethlehem, or the country of Judea. The text simply is not precise as to what sort of place it has in mind.

> And going into the house, they saw the child with Mary his mother, and they fell down and worshiped him. Then, opening their treasures, they offered him gifts, gold and frankincense and myrrh. (Matthew 2:11 ESV)

The text immediately transitions from the sighting of the star to going into the house. On this basis, it is often assumed that the place over which the star was standing was the house. However, the text is not explicit about this, and the text may be implying that they found the child by following Herod's instructions to search carefully for him.

It is clear that some of the ideas that Christians tend to have about the Magi's visit are not found in the text. Most notable is the idea that the Magi followed the star from their home country to Jerusalem. There is nothing in this text to suggest that and much which seems to conflict with it. Two further ideas derive from this first one. Firstly, it is sometimes claimed that the star was visible during the day. Nothing in the text indicates this, but it is necessary if the Magi were following the star during the day on their long journey. Secondly, it is sometimes claimed that the star disappeared when they got to Jerusalem, which seems to be an attempt to explain why they asked for the king of Jews instead of continuing to follow the star.

Several aspects of the story make more sense if we understand the star as a mundane astronomical event. The Magi were astrologers. The language describes an ordinary heliacal star rising. The visit to Jerusalem makes sense if the star was not guiding them. This understanding gives purpose to Herod's instructions to search diligently for the child.

Ultimately, our purpose is not to attempt to convince anyone that the Star of Bethlehem was merely a mundane astronomical event. It is arguable what the best interpretation of the brief account in the text is. However, it is important to appreciate that many common ideas about the Star of Bethlehem are legendary embellishments and are not found in the Biblical text.

If the Magi were astrologers, this causes a difficulty. If the Magi were able to find the Savior through using astrology, this would suggest that astrology works. The lack of any hint in the text of disapproval of the astrologers' practice might lead one to think that the author sees nothing wrong with astrology. It should be emphasized that in ancient astrology, the stars were not simply predicting or heralding events, they were causing them. As such, the historical Christians had to strongly reject the idea that the movements of the heavenly bodies had caused Christ's birth.

Historical Christians

Tertullian

Tertullian touched on this subject while writing against astrology.[1] He quotes those defending astrology saying, "But Magi and astrologers came from the east." He says that astrology was allowed until the time of the Gospel. Their gifts of gold, frankincense, and myrrh brought "worldly sacrifice and glory" to a close.

Tertullian argues that when the Magi were warned in a dream to return to their country by another way, this meant that they should cease to practice astrology. He insists that it was not to prevent Herod from pursuing them, because Herod did not know they went another way because Herod did not know which way they came by in the first place.

Origen

Origen discusses the Magi in response to the objection of Celsus.[2] Origen's first objection is that Celsus describes them as Chaldeans rather than Magi. The term Chaldean originally referred to a group of people in ancient Babylonia, but came to refer to astrologers and astronomers. Origen objects because he thinks that the Magi were not astrologers but rather sorcerers.

Origen states that the Star of Bethlehem was a new star unlike any of the known planets. It partook of the nature of a comet or a meteor. He argues that such stars often appear during the occurrence of great events. He gives the example of the removal of dynasties or the breaking out of wars. He cites a now lost work on comets by Chaeremon of Alexandria to show that positive events are sometimes heralded by comets as well.

---

[1] Tertullian, "On Idolatry," Chapter 9.
[2] Origen, "Contra Celsum," Book I, Chapter 58-60.

Given the momentous importance of the birth of Christ, what could be more fitting than such a new star?

According to Origen, the Magi were sorcerers who performed magic with the aid of evil spirits. However, at the birth of Christ, the spirits grew feeble. The Magi were thus attempting to determine why their magic had ceased to function. They conjectured that a being of great power must have arrived. At this time, they observed a new star in the heavens. They discovered the meaning behind this star in the words of Balaam:

> I see him, but not now; I behold him, but not near: a star shall come out of Jacob, and a scepter shall rise out of Israel; it shall crush the forehead of Moab and break down all the sons of Sheth. (Numbers 24:17 ESV)

From the reference to a star rising out of Jacob, the Magi determined that the prophesied messiah had come to Israel and thus set out to see him.

John Chrysostom

Chrysostom mentioned the star while discussing the Gospel of Matthew.[3] He notes that the "enemies of the truth" claimed that because a star appeared when Christ was born this indicated that astrology could be relied upon. It makes no sense for Jesus to be born according to the law of astrology but then to condemn it.

He asks what the Magi actually learned from this star. Did they learn that he was the King of the Jews? However, he was not a king. Furthermore, he would later say that his kingdom was not of this world. He never acted as a king, but lived in poverty. Even if they learned he was a king, why did they come? What sense does it make for these Magi to go worship some foreign king? What purpose would there be in worship-

---

[3] Chrysostom, "Homilies on Matthew," Homily 6.

ping an infant who would not appreciate or remember the homage paid to him? If his parents were rich or powerful, then showing homage to the infant might curry favor with them, but Jesus' parents were neither rich nor powerful.

Chrysostom also points out that the Magi's action is the reverse of typical astrological practice. Usually, an astrologist looks at the stars at the time of a person's birth and tries to predict their future. They do not look at the stars to indicate when someone has been born.

Chrysostom suggests that the star only appeared to be a star. Stars move from east to west, but this star moved from north to south, because "for so is Palestine situated with respect to Persia." This statement is somewhat odd because most of Persia is east, not north, of Israel. Nevertheless, the western-most part of Persian territory was north of Israel, which is probably what Chrysostom has in mind.

He gives a number of further reasons why the star could not be a normal star. It was visible during the day, outshining even the sun. However, it was not visible the entire time; it was visible on their journey to Jerusalem but hid itself until they were departing Jerusalem. Finally, the star could not have identified the very house where Jesus was staying unless it came down from its great height.

Ambrose of Milan

Ambrose discusses the Magi in a lengthy digression in his commentary on Luke.[4] He references an account that says that they are the descendants of Balaam. He describes the star as a new star, not one that had been in the heavens seen since

---

[4] Ambrose, *Commentary of Saint Ambrose on the Gospel According to Saint Luke,* Book II, Chapters 44-49.

creation. He suggests further that the Magi knew that this child represented the end of their magic arts.

Augustine of Hippo

Augustine discusses the Bethlehem star in a couple of places.[5] His primary concern is to argue that Jesus was not born because of the star, rather the star appeared because Jesus was born. Even if ordinary men's lives are governed by the stars, which Augustine rejects, that certainly is not the case for Jesus. He argues that the Magi only saw the star after Christ was born, and thus the star is not the cause of his birth. Rather, Christ caused the star as one of his many miracles.

On the star itself, Augustine indicates that it was not one of the regular stars present from the creation of the world. He refers to it as a new star, a new light, and a new constellation. He describes the star as leaving its course and coming down to the infant Christ. He states it "led the Magi to the exact spot where God the Word was born."

John of Damascus

John of Damascus comments on the star.[6] He is discussing comets, which he states are not among the stars made at creation, but are formed by divine command. Likewise, he thinks the same is true of the Bethlehem star. This is the case because the course of the star changed, sometimes moving from east to west and sometimes from north to south. At some times it was hidden and at other times it was revealed. The stars do not act this way, and thus it was not a regular star.

---

[5] Augustine, "Contra Faustum," Book II, Section V; Augustine, *Sermons 184-229Z, on the Liturgical Seasons*, Sermon 202.

[6] John of Damascus, "An Exposition of the Orthodox Faith," Book II, Chapter 7.

Thomas Aquinas

Aquinas considered whether the Bethlehem star was part of the heavenly system of stars.[7] He insists that it was clear that it was not one of the ordinary stars for many reasons. He cites both Augustine and Chrysostom for support. Firstly, no star moves from north to south, which is the direction from Persia to Judea. Secondly, the star was not only visible at night but also at midday. Thirdly, the star was sometimes visible and other times it was not. Fourthly, the star stopped when the Magi stopped, rather than moving continuously. Fifthly, the star came down to indicate the house where Jesus was to be found.

He discusses the words of the Magi, "We have seen His star in the east." He suggests that this means that the star was in the east when they first saw it, as opposed to the Magi's seeing the star while they were in the east.

Aquinas notes that some suggested that the star was the Holy Ghost or an angel. However, he finds it more likely that it was a newly created star, not in the heavens but in the air near the earth. This star was moved by divine will. He quotes a sermon by Pope Leo:

> A star of unusual brightness appeared to the three Magi in the east, which, through being more brilliant and more beautiful than the other stars, drew men's gaze and attention: so that they understood at once that such an unwonted event could not be devoid of purpose

Aquinas considers an objection that stars in the air are called comets and are considered signs of the deaths of kings, not their birth. In response, Aquinas states that it was not a comet because it was visible during the daytime and had varied

---

[7] Aquinas, *The Summa Theologiæ of St. Thomas Aquinas*, Part Three, Question 36, Article 7.

movements. Even so, Aquinas argues that it signified the destruction of the kingdoms of the earth.

John Calvin

Calvin addresses this issue in his commentary on Matthew 2:1–6.[8] He notes two questions: Was this one of the ordinary stars created in the beginning? Were the Magi led by astrology to conclude that it signified the birth of Christ? Calvin argues that it was not an ordinary star. It disappeared at times and suddenly became bright at others. It pursued a straight course for Bethlehem and remained stationary above Bethlehem for an extended period of time. Natural stars do none of these things. Calvin suggests that the light probably resembled a comet, but Matthew used popular language, calling it a star.

If the star was a supernatural star, astrology alone could not have guided the Magi to Christ. They may have received some assistance from astrology, but they were only able to reach Christ because they were aided by a new and extraordinary revelation. He suggests that they were guided by a secret revelation of the Spirit.

Conclusions

Historical Christians

Those authors who address the question conclude that the Star of Bethlehem is not one of the regular stars. Some mention meteors or comets. However, it should be kept in mind that they do not think of meteors and comets as natural phenomena like we do. Rather, they saw them as the direct action of God creating new stars.

---

[8] Calvin, *Calvin's Commentary on the Bible*.

Chrysostom, Aquinas, John of Damascus, and Calvin all give more detailed arguments for the star's not being a regular star. They claim that the star moved from north to south to lead the Magi to Jerusalem, that it was visible during the day, that it disappeared at times, and that it marked the exact house where Jesus was staying.

Some of the authors are primarily concerned to avoid the idea that this demonstrates the truth and morality of astrology. Tertullian suggests that it was the last allowed instance of astrology. Chrysostom insists that Jesus could not be born according the law of astrology. Augustine insists that the star did not cause Jesus' birth, but rather Jesus' birth caused the star.

Modern Science

Johannes Kepler, who played a key role in the development of heliocentrism, began a tradition of attempting to identify the Star of Bethlehem with a known astronomical phenomenon. It has been proposed the star was a simple star, a conjunction of planets, a comet, a meteor, or a supernova. Many proposals have passionate defenders but no proposal has gained broad acceptance.

Many Christians today would see the star as simply a miracle and be suspicious of attempts to identify it with any sort of natural phenomenon. However, understanding the event as a miracle raises the question of why no other sources reported it. One would expect that others would have reported this strange astronomical event if they did not know what it meant. But perhaps it was a light that could only be seen by these Magi.

Some skeptics of Christianity will argue that the story is simply an invented fable. They understand the text as describing a miraculous event that did not and could not happen. There is simply no way in which a star or similar body could move in the way described in the story. Even if it did, it would have been seen and widely reported.

Evaluation of the Response

Historical Christians respected the authority of the Biblical text. They assumed that if the Biblical text described a miraculously moving star, then there was a miraculously moving star. They may not have been sure how it fit into their understanding of astronomical phenomena, but they believed that the star had been present.

Many authors confused the legends that had been built up around the story with what is actually in the text. Indeed, the arguments put forward that the star is miraculous depend almost entirely on ideas that aren't in the text, like the star shining during the day, leading the Magi from their own country, and disappearing and reappearing. In some cases, they also sought to avoid the implication that the Magi were practicing astrology.

These authors were failing to return to the source. If we want to understand the nature of the Star of Bethlehem, we have to begin by going back to the source to see what the text really says. It is possible that the star was a relatively mundane astronomical event that God miraculously used to send the Magi to Jesus. Or perhaps there was a miraculous astronomical event somewhat different from stellar motions as typically imagined. Whatever the correct answer, we must go back to the source to find it.

Chapter 22

Do Some Living Things Arise by Spontaneous Generation?

The Scientific Question

Some living things clearly arise from similar living things. Sheep give birth to sheep. Birds lay eggs that grow into birds. However, in other cases, living things appear to arise out of nowhere. Maggots grow on dead flesh without any apparent adult maggots giving birth to them. Weeds grow just about anywhere without anyone's planting their seeds.

Classical science thought that some living things could arise spontaneously without the aid of a prior living thing. A wide variety of creatures, such as insects, worms, eels, frogs, salamanders, and fish, were thought to arise not from eggs or live birth but from mud, decaying flesh, soil, seaweed, etc. These living things formed naturally from these materials under nothing but the action of the sun's heat, wind, or similar phenomena.

The Biblical Discussion

There are a couple of verses in the creation account that speak of the earth's bringing forth living things:

> The earth brought forth vegetation, plants yielding seed according to their own kinds, and trees bearing fruit in which is their seed, each according to its kind. And God saw that it was good. (Genesis 1:12 ESV)

> And God said, "Let the earth bring forth living creatures according to their kinds—livestock and creeping things and beasts of the earth according to their kinds." And it was so. (Genesis 1:24 ESV)

The idea of the earth's bringing forth living things resembles the idea of spontaneous generation. However, the resemblance is rather slight. The Hebrew grammar indicates that the earth takes on a causal role in the emergence of these living things. But the earth does not play such a role in the ancient conception of spontaneous generation. The earth is not even the source material for many purported forms of spontaneous generation. The text is describing the origin of all living things, not simply those that were thought to arise by spontaneous generation. Nothing in the text indicates that this is an ongoing process rather than a special event that took place during creation. As such, the text cannot plausibly be referring to spontaneous generation.

On the other hand, a number of verses suggest an underlying assumption that all living things arise through reproduction. The creation account emphasizes that living things produce other living things. It emphasizes that the planets yield seed:

The earth brought forth vegetation, plants yielding seed according to their own kinds, and trees bearing fruit in which is their seed, each according to its kind. And God saw that it was good. (Genesis 1:12 ESV)

When creating birds and fish, God instructs them to be fruitful and multiply:

And God blessed them, saying, "Be fruitful and multiply and fill the waters in the seas, and let birds multiply on the earth." (Genesis 1:22 ESV)

The text emphasizes that God created a variety of kinds of living things:

The earth brought forth vegetation, plants yielding seed according to their own kinds, and trees bearing fruit in which is their seed, each according to its kind. And God saw that it was good. (Genesis 1:12 ESV)

And God blessed them, saying, "Be fruitful and multiply and fill the waters in the seas, and let birds multiply on the earth." (Genesis 1:22 ESV)

And God made the beasts of the earth according to their kinds and the livestock according to their kinds, and everything that creeps on the ground according to its kind. And God saw that it was good. (Genesis 1:25 ESV)

The flood account records that God had Noah preserve living specimens aboard the ark:

And of every living thing of all flesh, you shall bring two of every sort into the ark to keep them alive with you. They shall be male and female. Of the birds according to their kinds, and of the animals according to their kinds, of every creeping thing of the ground, according to its kind, two of

> every sort shall come in to you to keep them alive. (Genesis 6:19–20 ESV)

The implication is that if these living things had not been aboard the ark, they would have gone extinct.

While there are no explicit statements rejecting spontaneous generation, the creation and flood accounts make the most sense if spontaneous generation is false. It would make little sense for the text to emphasize their reproduction, describe God directly creating multiple kinds, and describe them as needing to be preserved aboard the ark if they arose by spontaneous generation.

Historical Christians

Origen

Origen mentions spontaneous generation while responding to Celsus.[1] Celsus argues that spontaneous generation shows that the physical bodies of different animals are all of the same nature. He draws the conclusion that only man's soul was created by God while his body was not.

Origen accepts spontaneous generation, referring to "laws established for the purpose of regulating the changes of bodies." He insists that it is not surprising that "a snake should be formed out of a dead man, growing, as the multitude affirm, out of the marrow of the back, and that a bee should spring from an ox, and a wasp from a horse, and a beetle from an ass, and, generally, worms from the most of bodies." However, Orgien argues that this does not mean that all bodies must have the same nature.

---

[1] Origen, "Contra Celsum," Book IV, Chapter 57.

Lactantiuss

Lactantius briefly mentions spontaneous generation while defending creation *ex nihilo*:[2]

> Nor, however, let it disturb any one, that certain animals appear to be born from the earth. For the earth does not give birth to these of itself, but the Spirit of God, without which nothing is produced.

His broader point is that animals have reason and thus must have been created by something with reason. However, if spontaneous generation were true, this would seem to show that some living things with reason arise without an intelligent creator. Lactantius insists that these animals are still born by the power of the Spirit of God and thus still have an intelligent creator.

Basil of Caesarea

Basil invokes spontaneous generation while commenting on the command for the earth to bring forth living creatures.[3] He acknowledges that certain creatures, such as horses, lions, and eagles, reproduce themselves. However, he thinks that other creatures are born from the earth itself. He mentions grasshoppers, insects so small they don't have names, mice, frogs, and eels. He interprets the command for the earth to bring forth living creatures as a reference to God's initiating spontaneous generation.

Augustine of Hippo

Augustine invokes spontaneous generation while arguing against the vegetarian stance of the Manichaeans.[4] In that

---

[2] Lactantius, "The Divine Institutes," Chapter IX.
[3] Basil of Caesarea, "Hexaemeron," Homily 9.
[4] Augustine, "On the Morals of the Manichaeans," Chapter 17.

philosophy, there were heavenly princes who were bound by the creator of the world to the physical world. Animals were descended from these princes. If one ate these animals, these princes would seek retribution for killing and eating their descendants. Augustine argues that Manichaeans ought to thus be willing to eat animals that arose by spontaneous generation since they do not descend from a heavenly prince, and yet this was not the practice of the Manichaeans.

Augustine suggests that those creatures that arose by spontaneous generation did not need to be aboard the ark.[5] He points specifically to the reference to male and female, suggesting that species without a distinction between male and female are not included.

Augustine discusses whether creatures produced by spontaneous generation were created during the initial creation or only arrived later.[6] He accepts, without question, that these creatures arise from excrement, the vapors of living bodies, the corruption of corpses, and decomposed plants. He insists that God is still the creator of these animals. He suggests that the creatures that arise from the waters and the earth were created along with the earth rather than with the later creation of the inhabitants of the earth. However, for those that arise from dead and decaying bodies, he insists that it would be absurd to say they were created along with those living creatures. Rather, God created a natural power for these living creatures to arise.

Thomas Aquinas

Aquinas considers spontaneous generation while discussing whether or not acts of creation occur during the ordinary

---

[5] Augustine, "The City of God," Book XV, Chapter 27.

[6] Augustine, *The Literal Meaning of Genesis. 1*, Book 3, Chapter 14.

operation of nature.[7] In most cases, like begets like. However, in the case of spontaneous generation, the thing formed is unlike that which it is formed from. This would seem to be an act of creation. However, Aquinas has argued that creation is always from nothing and that only God can create. Aquinas attributes the formation of these creatures to a celestial power, that is, the action of the stars. The form imparted to the generated creatures is part of the form of the stars' acting on the creatures, not a true act of creation.

Martin Luther

Luther discusses the origins of mice and lice while commenting on Genesis 1:21.[8] He points out that even ships that are continually on the sea have mice in them. No house is so clean that mice will not be found in it. He mentions Aristotle, who taught that while some animals are generated by ordinary reproduction, other animals are produced both by ordinary reproduction and spontaneous generation. According to Aristotle, this happens as a result of the warming effect of the sun.

Luther, however, expresses skepticism. The sun warms everything, but nothing is generated unless God speaks it into being. Even if it were true that mice arose from putrid matter, it could only be because of the action of God's divine power. However, Luther argues that if mice were created by spontaneous generation, they would not have a consistent form. Rather, because mice have a beautiful form with exquisite symmetry and smooth hair, they must be a particular order of creatures created by God. The same is true, he suggests, of flies. In order to explain how mice get onto ships, Luther

---

[7] Aquinas, *The Summa Theologiæ of St. Thomas Aquinas*, Part 1, Question 45, Article 8, Objection 3.

[8] Luther, *Luther On The Creation: A Critical and Devotional Commentry on Genesis.*

suggests that the mouse is of an "aqueous nature." He does not elaborate on what he means by this.

John Calvin

In his Commentary on Genesis 1:20,[9] Calvin interprets the text as indicating that innumerable fish are produced from the waters every day. In his translation, the text refers to waters' bringing forth sea creatures. Most modern translations instead indicate that the sea is to be filled with living creatures, not bring them forth. He insists that the language of Moses expresses not only an initial creation but the continual production of sea creatures from the waters.

Conclusions

Historical Christians

Historical Christian authors generally unquestioningly accepted spontaneous generation. Sometimes, they noticed tension between Christian beliefs and spontaneous generation but the authors came up with an explanation that reconciled spontaneous generation and Christianity. Some others explicitly identified the creation account as referring to spontaneous generation. Luther alone, among these authors, expressed skepticism towards spontaneous generation.

Modern Science

Jan Swammerdam, a Dutch biologist, is notable for being one of the first biologists to object to spontaneous generation. He thought it impious. In 1669, he published a book entitled *The General History of Insects,* in which he documented the complex

---

[9] Calvin, *Calvin's Commentary on the Bible.*

internal structure of insects and argued they were produced by tiny eggs and not spontaneous generation. In 1668, Francesco Redi performed a careful experiment showing that maggots did not spontaneously arise from rotting meat. When he carefully ensured that the flies could not access the meat to lay their eggs, no maggots formed on the meat. It became increasingly clear that spontaneous generation was not a real phenomenon. The debate culminated with Louis Pasteur's experiments in 1859. These experiments are typically credited with settling the issue, demonstrating that spontaneous generation did not exist even at the level of microbes.

Evaluation of the Response

Most historical Christians accepted spontaneous generation without question. Luther alone expressed skepticism about the idea. In this, they were following the science of their day. Some went so far as to argue that the creation account referred to spontaneous generation. But spontaneous generation was an incorrect theory.

It is not surprising that historical Christians did not question spontaneous generation. It was such a widely accepted idea that it would never have occurred to most people to question it. Furthermore, the Bible does not contain any passages that directly contradict the idea. A Biblical case against spontaneous generation can be made, but it is subtle.

Even so, some authors did notice a tension between spontaneous creation and the doctrine of creation. However, they did not take this as an reason to question spontaneous generation. Rather, they assumed that spontaneous generation was true and adjusted their theology to fit. In doing so, they failed to return to the source. Eventually, Christian scientists did return to the source, question spontaneous generation, and determine that spontaneous generation was not real.

Chapter 23

Was Hebrew the First Language?

The Biblical Issue

The story of the Tower of Babel begins by declaring that a single language was used all over the earth:

> Now the whole earth had one language and the same words. (Genesis 11:1 ESV)

The text goes on to explain the origin of various languages as a judgment against man because of the building of the Tower of Babel. But what was the original language? What language did everyone on earth speak before the languages were confused and divided? A longstanding idea among Biblical commentators is that the language in question was Hebrew. On this understanding, the line leading to Abraham preserved the original human language, whereas the other lines had their language confused.

One reason for this is that the text records God's assigning names during creation:

> God called the light Day, and the darkness he called Night. And there was evening and there was morning, the first day. (Genesis 1:5 ESV)

And God called the expanse Heaven. And there was evening and there was morning, the second day. (Genesis 1:8 ESV)

God called the dry land Earth, and the waters that were gathered together he called Seas. And God saw that it was good. (Genesis 1:10 ESV)

The names recorded here are Hebrew names, יוֹם (*yom*) meaning day, לַיִל (*layil*) meaning night, שָׁמַיִם (*shmayim*) meaning heavens, אֶרֶץ (*erets*) meaning earth, and יָם (*yam*) meaning seas.

Later, Adam gives names to the animals:

Now out of the ground the LORD God had formed every beast of the field and every bird of the heavens and brought them to the man to see what he would call them. And whatever the man called every living creature, that was its name. The man gave names to all livestock and to the birds of the heavens and to every beast of the field. But for Adam there was not found a helper fit for him. (Genesis 2:19–20 ESV)

After the creation of Eve, Adam gives her the name woman:

Then the man said, "This at last is bone of my bones and flesh of my flesh; she shall be called Woman, because she was taken out of Man." (Genesis 2:23 ESV)

Later, he gives her the name Eve:

The man called his wife's name Eve, because she was the mother of all living. (Genesis 3:20 ESV)

The recorded names are in Hebrew, אִשָּׁה (*issah*) meaning woman and חַוָּה (*chavvah*) which we render as Eve. One might assume that the names being discussed are in Hebrew because it is Adam's Hebrew names that are assigned in the text.

However, the Bible uses Hebrew names throughout, even when discussing foreign entities. It does not refer to Assyria, Babylon, Egypt, or Ethiopia, but rather to Asshur, Babel,

Mitsrayim, and Cush. The Septuagint often, but not always, translated names using the Greek equivalents. Thus, while the Hebrew text would say, Mitsrayim, the Septuagint would say Αἴγυπτος (*aigyptos*). This practice is followed by many modern translations, which give the impression that certain names are Hebrew while others are not. In fact, the text always uses Hebrew names. As such, the fact that the names are in Hebrew in the text does not necessarily indicate that they were Hebrew names originally.

However, the text also makes puns on the names. The Hebrew word for ground is אֲדָמָה (*adamah*), and Adam is made from the dust of the ground. Adam calls his wife אִשָּׁה (*issah*) because she was taken out of אִישׁ (*ish*) him. Later, he calls his wife חַוָּה (*chavvah*), because she is the mother of all living. The Hebrew word for living is חַי (*chay*). Puns, by their nature, do not translate. As such, it may be argued that the language they used must have been Hebrew in order for the puns to work.

On the other hand, there is nothing unusual about a Hebrew-speaking author engaging in Hebrew wordplay. Some of the puns might be explained as the author's choosing to put things in a way that emphasized Hebrew wordplay. Furthermore, puns can sometimes be translated. For example, English translations have Adam saying that Eve will be "called women for she came out of man," giving a similar pun as the Hebrew text. As such, the presence of Hebrew wordplay may be explained without the original language's also being Hebrew.

Conspicuously absent from the Tower of Babel story is any indication that the original language was maintained by any group. The story does not indicate there was some righteous group that was not subjected to the general punishment. Furthermore, the Biblical indication is that the Jews' ancestors worshipped other gods:

And Joshua said to all the people, "Thus says the LORD, the God of Israel, 'Long ago, your fathers lived beyond the Euphrates, Terah, the father of Abraham and of Nahor; and they served other gods.'" (Joshua 24:2 ESV)

Genesis 10 contains a list of descendants of Noah. It is often called the table of nations, because it lists nations descended from the three sons of Noah. Further, it describes each of these nations as having its own language:

From these the coastland peoples spread in their lands, each with his own language, by their clans, in their nations. (Genesis 10:5 ESV)

One notable name in that list is Eber, also sometimes spelled Heber. In Hebrew, the name is עֵבֶר (*eber*). He was a great-grandson of Shem and a distant ancestor of Abraham. Our word Hebrew derives from the Hebrew word עִבְרִי (*ibri*). Christians have long thought that Eber was, in fact, the namesake of the Hebrews. Many thought he was a righteous man who maintained the original language, which became known as Hebrew. Modern scholars typically favor the interpretation that עִבְרִי (*ibri*) derives from a word meaning beyond or across. The name is thus understood as a reference to their being immigrants from across the river.

However, there is reason to think that Hebrew is not the language of Eber or even Abraham. Rather, it was the language spoken by the Canaanites. Abraham or his descendants adopted the language after immigrating to Canaan. Isaiah appears to refer to it as the language of Canaan:

In that day there will be five cities in the land of Egypt that speak the language of Canaan and swear allegiance to the LORD of hosts. One of these will be called the City of Destruction. (Isaiah 19:18 ESV)

Furthermore, throughout the interaction of Abraham's descendants and the people living in Canaan, there is no indication of any language barrier. Rather, in many instances, Israelites freely communicate with their non-Israelite neighbors. In contrast, Genesis 42:23 indicates that Jacob's sons used an interpreter to speak to Egyptians. In 2 Kings 18, 2 Chronicles 32, and Isaiah 36, Sennacherib's messenger makes a point of speaking in the language of Judah so everyone hearing him can understand his threats.

The language is never identified with Abraham, Eber, or even the term Ibri. It is identified in this passage in Isaiah as the language of Canaan, and in some other places as the language of Judah. As such, the indications we have about the language do not suggest it is a language deriving from Abraham but rather is the common language of the land of Canaan.

It is clear that there is no Biblical statement that would indicate that Hebrew was the original language. The presence of Hebrew names does not mean that the names are not translated from a different language. Furthermore, a straightforward reading of the Tower of Babel story gives no indication of any exemption. The text rather suggests that Hebrew is simply the language already spoken in the land of Canaan. Nevertheless, the Hebrew wordplay gives some limited reason to think the original language might have been Hebrew.

Historical Christians

Origen

Origen discusses that the original language and the division of languages.[1] He argues that God sent angels to confuse

---

[1] Origen and Scheck, *Homilies on Numbers*, Homily 11, Section 4. 4; Origen, "Contra Celsum," Book 5, Chapter 30-31.

the languages at the Tower of Babel with one angel inspiring the Babylonian language, another inspiring the Egyptian language, and still another inspiring the Greek language. However, the original language was preserved in the part of humanity that remained God's portion because they did not participate in the building of the tower.

John Chrysostom

Chrysostom refers to the names Adam gave the animals.[2] He states these names "remain up to the present time." Furthermore, they serve as a constant reminder of the esteem which humans received from the Lord. In Homily 30, he said that Eber kept the original language, implying that he thinks that Hebrew was the original language.

There is something of a conflict between Chrysostom's belief that Hebrew was the original language and his belief that the names Adam gave the animals were the same ones in use by his hearers. It may be that Chrysostom simply had not considered that the names given to the animals would have been in Hebrew. Alternatively, perhaps he considers the Greek names he would have used as translations of Adam's Hebrew names.

Gregory of Nyssa

Gregory touches on this issue while responding to Eunomius, who held to an extreme form of Arianism.[3] A major point in dispute involved the name or term, $A\gamma\acute{\epsilon}\nu\nu\eta\tau o\varsigma$ (agennētos), which means unbeginning. Eunomius insists that this was a name that God gave himself. However, Gregory argues that it was a name given to God by man based on man's understand-

---

[2] Chrysostom, *Homilies on Genesis*, Homily 14.
[3] Gregory of Nyssa, "Answer to Eunomius' Second Book."

ing of God's nature. Eunomius also claimed that God named "germ, and herb, and grass, and seed, and tree, and the like" when he created them. Gregory insists that these names were given not by God but by man.

Gregory argues that God could not have named these things because that would mean he would have had to speak, pronouncing the sounds making up these names. He could not speak without having a physical body with vocal organs. However, the Biblical text does describe God as speaking. As the famous quotation goes, "God said, 'Let there be light.'" Gregory argues that the text cannot intend to mean that God literally spoke, because there was nobody to listen. God, he insists, is not ordering himself around. He anticipates an objection that the Father might be speaking to the other members of the Trinity; however, Gregory insists that the members of the Trinity would not require audible noise in order to communicate.

Gregory cites the psalmist:

> The heavens declare the glory of God, and the sky above proclaims his handiwork. Day to day pours out speech, and night to night reveals knowledge. There is no speech, nor are there words, whose voice is not heard. (Psalm 19:1b–3 ESV)

The psalmist speaks of the heavens' declaring and pouring out speech, yet there is neither speech nor words, and their voice is not heard. Gregory argues that the psalmist was using a bodily illustration. The heavens make clear God's glory without engaging in physical speech. Gregory argues that something similar is occurring in the Genesis creation account. Moses is describing God's creating the world by his will, using the metaphor of his speaking. Gregory argues that if one insists on taking "God said" to mean that God spoke audibly, one must also take all of the other passages of the Bible that speak of God anthropomorphically in the same way.

Gregory argues that if God had given names to the elements of creation, those names would be common to all men. God made the nature of the elements, and that nature appears the same to all men. Likewise, Gregory believes that if God had imposed names on those elements, all men would have had the same names. However, it is clear that different languages use different names for the same elements. Gregory anticipates an appeal to the Tower of Babel in response. He insists that even in that account, God did not create language but simply confused the languages that were already present.

Initially, Gregory describes the language spoken by Moses as one of these subsequent languages. However, he then suggests that "some who have carefully studied Scriptures" said that Hebrew was not one of these languages produced at the Tower of Babel. Instead, that language was miraculously formed at the time of the Exodus. If Hebrew is indeed a much later language, it would make no sense to suggest that God was speaking and naming things in that language.

To support the idea that Hebrew was miraculously formed at the time of Exodus, he cites the psalmist:

> For it is a statute for Israel, a rule of the God of Jacob. He made it a decree in Joseph when he went out over the land of Egypt. I hear a language I had not known: (Psalm 81:4–5 ESV)

Gregory takes reference to the unknown language as indicating that Hebrew was a new language nobody had previously known. Most other commentators think the reference is to the Egyptian language, which the Hebrews did not understand.

Nevertheless, the text describes God as naming earth, heavens, day, and night. Gregory argues that God is not assigning names but is creating one thing from another. Thus, he argues that God created day from light, but not did give the name light to day. He argues similarly for the various names assigned by God.

Augustine of Hippo

Augustine writes that the original language survived in the descendants of Eber.[4] In this work, Augustine is tracing the line of the righteous followers of the true God, which he calls the City of God. He acknowledges that the text does not identify any individuals as still worshipping the true God between Noah's sons and Abraham. However, Augustine says that he "cannot believe that there actually were none." He insists that there was never a time when there were not both those who worshipped the true God and those who held him in contempt.

In support of this idea, Augustine cites the psalmist:

> The LORD looks down from heaven on the children of man, to see if there are any who understand, who seek after God. They have all turned aside; together they have become corrupt; there is none who does good, not even one. Have they no knowledge, all the evildoers who eat up my people as they eat bread and do not call upon the LORD? (Psalm 14:2–4 ESV)

The first couple of verses seem to describe a situation where the Lord looks down from heaven and finds nobody who seeks after him. Nevertheless, the last verse makes it clear that there are still people whom God identifies as "my people." Therefore, there are still people who follow God even when it seems there are none.

---

[4] Augustine, "The City of God," Book XVI.

He points out that the text draws attention to Eber, mentioning him immediately after Shem, even though he was born five generations later:

> To Shem also, the father of all the children of Eber, the elder brother of Japheth, children were born. The sons of Shem: Elam, Asshur, Arpachshad, Lud, and Aram. The sons of Aram: Uz, Hul, Gether, and Mash. Arpachshad fathered Shelah; and Shelah fathered Eber. (Genesis 10:21–24 ESV)

Augustine argues that the reason this attention is given to Eber is because he is the namesake of the Hebrews and the Hebrew language. He identifies Eber as the representative of the line that still followed God, and thus maintained the original human language when the languages were divided. It became known as Hebrew because it was the language of Eber.

However, Augustine's idea faces a difficulty in Eber's son Peleg:

> To Eber were born two sons: the name of the one was Peleg, for in his days the earth was divided, and his brother's name was Joktan. (Genesis 10:25 ESV)

Peleg is related to the word meaning to divide. The days in which the earth was divided is typically taken to be reference to the division of languages. But if the languages were divided in Peleg's day, it would seem that they were not divided in the day of Eber, and thus Eber would not be the namesake of Hebrew. Augustine answers that the division happened when Peleg was born and this prompted Eber to use the name Peleg.

Genesis 10 has traditionally been taken as identifying the origin of seventy nations. However, obtaining the number seventy requires not counting Nimrod, whose exploits are recounted in verses 9–12, while counting the fathers in the genealogy as founding distinct nations from their sons. Augustine was using a translation derived from the Septuagint, which adds three names and removes one, and thus he considers there to be seventy-two nations listed.

But this causes a problem for him, because if Eber named Peleg at around the time of division of languages, Peleg would not have founded his own nation with its own language. Augustine is unwilling to go against the tradition of identifying seventy or seventy-two nations in this text. He resolves this by arguing that Nimrod should be counted as founding his own nation, replacing Peleg in the count. He argues that Peleg is included in the list because of the timing of his birth, not because he founded a nation.

However, Nimrod is four generations removed from Noah, while Eber is six generations removed. Nimrod is described as being the founder of Babel. But this would suggest that Eber lived two generations after Nimrod, not at the same time at the Tower of Babel. Augustine responds that some generations are shorter than others. Furthermore, by the time of division of languages, the various families must have been large enough to be able to become peoples upon the division of the languages.

Remigius of Auxerre

Remigius of Auxerre discussed the language that Adam used in naming the animals.[5] He says that there is "little doubt that it was Hebrew." As evidence, he states that all the names of people or places in Genesis are in Hebrew, up until the Tower of Babel. He makes an exception for fish that were not brought to Adam to name, and he indicates that different nations gave names to different sorts of fish gradually over time.

---

[5] Remigius of Auxerre, "Exposition on Genesis," Chapter Two.

Bede

Bede addresses the issue in his commentary on Genesis.[6] He gives the familiar argument that the language was Hebrew because the names in the Bible are in Hebrew up until the time of the division of languages.

Dante

Dante discusses the original language.[7] He notes that everyone may be inclined to think their own language was the original language. However, Dante insists that he will not be swayed by such sentiments. Rather, he says the original language was spoken until the Tower of Babel and then inherited by the sons of Eber, whom the Hebrews are named after. Thus, our Redeemer, Jesus Christ, spoke the language not of the confusion wrought by Babel but of grace.

However, Dante gives a different account in his epic poem, *The Divine Comedy*.[8] The Divine Comedy is a story in which Dante travels through hell, purgatory, and heaven. In this part of the poem, he meets Adam in heaven. The character Adam declares that his language was extinct before the Tower of Babel. He notes instead that language changes over time. Adam says that the word for God was originally *"I"* but became *"El."* *El* is one of the Hebrew words for God.

Conclusions

Historical Christians

There is a strong tradition of holding that Hebrew was the original language. Origen, Chrysostom, Augustine, Remigius,

---

[6] Bede, "Commentary on Genesis."

[7] Dante Alighieri, *De Vulgari Eloquentia*.

[8] Dante Alighieri, *The Divine Comedy*, Paradise, Canto 26.

and Bede all took this position. Gregory of Nyssa gives a vigorous dissent, arguing that Hebrew isn't even an ancient language. Dante appears to give different answers in different places. There are two common themes put forward to argue for this idea. The first is that the names in the text are in Hebrew up to the time of the Tower of Babel. The second is that Eber is the namesake of the Hebrews.

Modern Science

The late eighteenth century saw the development of historical linguistics, the study of the development of human languages. It became clear that languages change over time. Furthermore, languages diverged, with one language becoming multiple languages over time. This suggested the possibility that there was one human language that had, over time, been divided and changed into the numerous languages known today.

Linguists would go on to identify groups of languages that exhibited similarities. Languages as diverse as Greek, Latin, English, Russian, and Hindi are classified Indo-European languages and are thought to derive from a single original proto-European language. Hebrew is an Afro-Asiatic language. It appears, like all other languages, to be yet another language derived by the ordinary processes of language evolution. It is in fact a dialect of Canaanite, suggesting that it was indeed the language of Canaan. Abraham would have likely spoken another language in his homeland of Ur, but his descendants adopted the language of Canaan eventually creating their own dialect.

Evaluation of the Response

Almost all historical Christians who addressed this subject held that Hebrew was the original language and had been maintained by Eber and the Israelites. Such a position would be extremely difficult to defend today. It rather appears clear that Hebrew is a dialect of the Canaanite language.

The historical Christian belief that Hebrew was the original language was based on a poor understanding of the Hebrew language and text. Historical Christians point out that the text uses Hebrew names before the Tower of Babel. However, the Hebrew text always uses Hebrew names. It is translations like the Septuagint that translate the names inconsistently. They thought that Eber was the namesake of the Hebrews. However, there is no Biblical indication that this is the case. They failed to go back to the source and were led astray by translations.

Chapter 24

Do Humans All Come from the Same Place?

The Scientific Question

People from different parts of the world look different. For example, many people whose ancestors lived in Africa have darker skin. People from the Netherlands or Iceland often have blue eyes. Asians often have a distinctive eye shape that is rare among Europeans. They also speak different languages and have different customs.

Why these differences? A simple explanation would be that different races with different appearances and customs have distinct origins. The gods may have created different races at different times in different places with different appearances and customs. Different distinct groups of humans may have existed into an eternal past. Various different races may have emerged via a form of spontaneous generation. Whatever the case, different groups of people have separate origins, thus explaining their distinctive appearances and customs.

The alternative explanation is that there was only one origin of humanity but humans later diverged into the numerous races that we know today. In this scenario, humans must have begun with a small group in one location. The original humans might either have been created or have arisen in some other way. Over time these humans spread out across the world, diverging in their customs, appearance, and language to give rise to the races that we know today.

The Biblical Discussion

The Bible contains many statements that indicate that Adam and Eve were the original humans and that all humans descend from them. The text indicates that prior to the creation of Adam, there were no human beings in existence.

> When no bush of the field was yet in the land and no small plant of the field had yet sprung up—for the LORD God had not caused it to rain on the land, and there was no man to work the ground, (Genesis 2:5 ESV)

No helper fit for Adam was found, necessitating the creation of Eve.

> The man gave names to all livestock and to the birds of the heavens and to every beast of the field. But for Adam there was not found a helper fit for him. (Genesis 2:20 ESV)

The text indicates that Eve was the mother of all living human beings:

> The man called his wife's name Eve, because she was the mother of all living. (Genesis 3:20 ESV)

Two verses state that the people of earth descend from Noah's three sons:

> These three were the sons of Noah, and from these the people of the whole earth were dispersed. (Genesis 9:19 ESV)

> These are the clans of the sons of Noah, according to their genealogies, in their nations, and from these the nations spread abroad on the earth after the flood. (Genesis 10:32 ESV)

Paul, speaking at the Areopagus, states that God made every nation from one man:

> And he made from one man every nation of mankind to live on all the face of the earth, having determined allotted periods and the boundaries of their dwelling place, (Acts 17:26 ESV)

The Biblical text indicates that all humans descend from Adam and Eve as well as Noah's three sons. Different races do not have distinct origins.

However, this leaves a question. Where did the first generation of human children find spouses? Who did Cain and Seth marry? If all humans descended from Adam and Eve, the inescapable conclusion is that the first generation of men must have married their sisters. In the early generations, humans married very close relatives until the population increased enough that more distant relatives were available.

However, marrying one's own sibling is incest and violates a very strong and nearly universal taboo as well as Biblical law:

> You shall not uncover the nakedness of your sister, your father's daughter or your mother's daughter, whether brought up in the family or in another home. (Leviticus 18:9 ESV)

Nevertheless, such marriages did take place. Abraham was married to his half-sister, a relationship later forbidden by this Mosaic law:

> Besides, she is indeed my sister, the daughter of my father though not the daughter of my mother, and she became my wife. (Genesis 20:12 ESV)

As such, given no other options, it seems plausible that early humans married their siblings.

Historical Christians

Hippolytus of Rome

Hippolytus states that the nations of the world descend from the three sons of Noah.[1] His overall argument is that Christianity represented the original religion of mankind. The philosophies and beliefs of other nations, such as the Greeks, were corruptions of the original belief in the one true God. He goes on to point out that even if these nations were inhabited by people who did not descend from Noah, the flood story still shows that the worship of the true God predates all other belief systems.

Theophilus of Antioch

Theophilus discusses the flood, comparing the Biblical account to that found in Greek myths.[2] He mentions Plato as saying that the flood only covered the plains and that those who fled to the highest hills were spared. He also mentioned the myth of Deucalion and Pyrrha, who threw stones behind them, which became men and women to repopulate the earth. He also mentions Clymenus, who was associated with a second flood. In contrast to all of these stories, Theophilus insists that Moses got it right. Only those on the ark survived to repopulate the earth.

---

[1] Hippolytus, "The Refutation of All Heresies," Chapter XXVIII.
[2] Theophilus of Antioch, "To Autolycus," Chapter 18-19.

Origen of Alexandria

Origen mentions this issue while responding to Celsus's criticism of the flood account.[3] He declares that God arranged that all the seeds of living things would be brought into the ark and brought again out into the earth. He also speaks of a most righteous man who became the progenitor of all who were born after the flood.

John Chrysostom

Chrysostom discusses the question of Cain's wife.[4] He says, "But perhaps someone may say: How is it that Cain had a wife when Sacred Scripture nowhere makes mention of another woman?" He says that Scripture avoids superfluous details, mentioning only males and not even all of them. He argues that it is likely that Eve gave birth to a daughter after Cain and Abel and that Cain took this daughter to be his wife. He explains that it was permissible in the beginning for the early humans to marry their own sisters because "the human race had to increase from then on."

Augustine

Augustine discusses the origin of Cain's wife.[5] He wrote that because there were no human beings except those born to Adam and Eve, early men must have taken their own sisters as wives. This was "certainly dictated by necessity in these ancient days, as afterwards it was condemned by the prohibitions of religion."

Augustine wrote that it is better for people to have relationships with a wider variety of people. Thus, it is better that the

---

[3] Origen, "Contra Celsum," Book IV, Chapter 41.
[4] Chrysostom, *Homilies on Genesis*, Homily 20.
[5] Augustine, "The City of God," Book XV, Chapter 16.

same woman not be both a sister and a wife to her husband. This was not possible in the early days of the human race, but once it became possible, it became undesirable, even abominable, to marry one's own sister. Indeed, even though such marriages were legal in some places, people shrank from them.

He goes on to discuss marriages between cousins. He notes that while it was lawful, people shrank from it as being too similar to marrying a sibling. He sees this as a further development of the taboo against sibling marriages.

Later, Augustine discusses the dispersion of animals after the flood.[6] How did they reach the islands? He assumes that some animals, like frogs, were spontaneously generated and simply were spontaneously generated again after the flood. However, other animals, such as wolves, arose by reproduction and thus must have come from the flood. He suggests that some might have reached the islands by swimming, but this explanation could only work for islands near the mainland. Man may have transported some animals, but there does not seem to be a compelling reason for them to do this. He also suggests that they might have been transported to the islands by angels. Finally, he suggests that these animals may have been created again out of the earth as at the first creation. In this case, the animals were taken aboard the ark, not to preserve their kind but to prefigure the saving of the nations through the church.

In the next chapter, Augustine considers whether monstrous races were descended from Adam and Noah.[7] He speaks of reports of races of men with one eye, with backward feet, with both male and female features, lacking mouths, of extremely small stature, with short lifespans, without heads,

---

[6] Augustine, Book XVI, Chapter 7.
[7] Augustine, Book XVI, Chapter 8.

or with a single leg with two feet. Augustine notes that we do not need to believe all of these accounts. Furthermore, some of these might be non-human creatures.

Nevertheless, he insists that all rational moral animals are men, no matter how strange or peculiar. He insists that no Christian can doubt that they must spring from Adam's stock. To account for the possibility of such races, Augustine appeals to examples of birth deformities, conjoined twins, and intersex individuals. People with unusual characteristics are sometimes born to those who lack those characteristics. If these people had given rise to a race sharing their unusualness, they would seem as monstrous as these reported races.

Next, Augustine speaks of the Antipodeans, men thought to live on theoppposite side of the round earth.[8] He rejects this as "on no ground credible." He points out that even if it is proved that the earth is round, this does not show that the other side is not covered by water or people. It is absurd, Augustine insists, to think that men might have set sail from his side of the world to the other.

Martin Luther

Luther discusses the identity of Cain's wife.[9] He notes uncertainty as to whether Cain was married before or after murdering his brother. He also notes uncertainty regarding a Jewish story in which Eve had male and female twins at each birth. However, he says that it is certain that Cain married his sister. He thinks that the text indicates that Cain and Abel had both received their inheritance, with the care of cattle given to Abel by Adam and the tilling of ground given to Cain. Because of

---

[8] Augustine, Book XVI, Chapter 8.

[9] Luther, *Luther On The Creation: A Critical and Devotional Commentry on Genesis*, 1:Volume II, Section VII.

this, he thinks that Cain was probably already married when he killed Abel.

John Calvin

Calvin discusses Cain's wife.[10] He concludes that Cain married his sister before murdering his brother. He argues that if any of Cain's sisters had been willing to marry him after he murdered Abel, this would be a fact that Moses would have taken the time to record. When commenting on God's message to Noah after the flood,[11] Calvin writes that Noah and his progeny would diffuse over the earth and be responsible for repopulating it.

Conclusions

Historical Christians

Historically, Christians were consistent in their views. Adam and Eve were the original humans. All humans who ever lived descended from them. Cain, Seth, and any other brothers married their sisters. The only survivors of the flood were aboard the ark, and all nations of the world today descend from Noah's three sons. They held to this picture of history despite criticism of it that was sometimes mounted by critics.

Modern Science

The idea that different races of humans had distinct origins enjoyed a resurgence in the age of discovery as far-flung races were discovered around the world. However, mainstream modern science has rejected any such theory in favor of the

---

[10] Calvin, *Calvin's Commentary on the Bible*, Commentary on Genesis 4.
[11] Calvin, Commentary on Genesis 9:1.

evolutionary origin of humanity. According to the most widely accepted model, modern humans originated in eastern Africa and spread across the world from there. All cultures, languages, races, and peoples diverged from that original group of humans.

However, while mainstream modern theory agrees with the Biblical account in terms of a single population diverging and spreading over the earth, this does not mean that modern science accepts the Biblical Adam and Eve. In an evolutionary scenario, there was no original founding couple for the human race. Rather, a larger population gradually became humans through a process of evolution. That population was quite small at one time, perhaps as small as a thousand people, but never as small as just two people.[12]

Furthermore, under evolutionary theory, not only do all human races share common ancestry, humans share common ancestry with all living things. All living things, from bacteria to plants to animals and even humans, have diverged from a single origin of life in the distant past. Additionally, the time required for the human race to spread around the earth is thought to be much longer under evolutionary theory than the several thousand years given by the traditional Christian chronology.

Evaluation of the Response

Modern science has confirmed the Christian claim that the human race originated from a single population. However, the details regarding how small that population was, where it came from, and how long it took for that population to spread across the earth are very different. The question of how Chris-

---

[12] Huff et al., "Mobile Elements Reveal Small Population Size in the Ancient Ancestors of Homo Sapiens"; Hu et al., "Genomic Inference of a Severe Human Bottleneck during the Early to Middle Pleistocene Transition."

tians should respond to modern evolutionary theory is one of the most controversial questions in the Church today. But what can we learn from the historical debate?

We do see another example of historical Christians respecting the authority of the Biblical text. They accepted the Biblical account: Adam and Eve were the progenitors of the human race; only the occupants of the ark survived the flood; and all the nations of the earth descended from Noah's three sons. Despite some objections, none of these authors questioned the veracity of the text.

There is also encouragement in the fact that part of the Biblical narrative has been confirmed. We should not downplay the difference between Biblical narrative and evolutionary theory, but it is notable that evolutionary theory agrees with the Biblical account that humans come from a single population that spread across the earth. This is a case where the modern scientific idea is closer to the Biblical view than the ancient scientific idea.

Chapter 25

Did the Antediluvians Live for Centuries?

Genesis 5 contains a genealogy tracing from Adam to Noah. Each entry is similar to the following:

> When Seth had lived 105 years, he fathered Enosh. Seth lived after he fathered Enosh 807 years and had other sons and daughters. Thus all the days of Seth were 912 years, and he died. (Genesis 5:6–8 ESV).

Genesis 11 has similar entries for individuals living after the flood.

> When Arpachshad had lived 35 years, he fathered Shelah. And Arpachshad lived after he fathered Shelah 403 years and had other sons and daughters. (Genesis 11:12–13 ESV)

The patriarchs of Israel also have long lives attributed to them:

> These are the days of the years of Abraham's life, 175 years. (Genesis 25:7 ESV)

> Now the days of Isaac were 180 years. (Genesis 35:28 ESV)

> And Jacob said to Pharaoh, "The days of the years of my sojourning are 130 years. Few and evil have been the days of the years of my life, and they have not attained to the days

of the years of the life of my fathers in the days of their sojourning." (Genesis 47:9 ESV)

And Jacob lived in the land of Egypt seventeen years. So the days of Jacob, the years of his life, were 147 years. (Genesis 47:28 ESV)

These passages seem to describe lifespans far exceeding those of either modern or ancient experience. A straightforward reading is that the Biblical text claims that people really did live for a much longer time in the past. They lived for centuries prior to the flood. Immediately after the flood, their lives were shorter but still much longer than our own.

However, Abraham considered himself too old to have children at merely one hundred years old:

Then Abraham fell on his face and laughed and said to himself, "Shall a child be born to a man who is a hundred years old? Shall Sarah, who is ninety years old, bear a child?" (Genesis 17:17 ESV)

If people had lived for centuries and many had children when they were well past one hundred, why does Abraham find it so incredible that he could have a child at his age?

However, it had been many generations since anyone had fathered a child at such at advanced age. Consider the genealogy leading to Abraham:

When Arpachshad had lived 35 years, he fathered Shelah. And Arpachshad lived after he fathered Shelah 403 years and had other sons and daughters. When Shelah had lived 30 years, he fathered Eber. And Shelah lived after he fathered Eber 403 years and had other sons and daughters. When Eber had lived 34 years, he fathered Peleg. And Eber lived after he fathered Peleg 430 years and had other sons and daughters. When Peleg had lived 30 years, he fathered Reu. And Peleg lived after he fathered Reu 209 years and had other sons and daughters. When Reu had lived 32 years, he fathered Serug. And Reu lived after he fathered Serug 207 years and had other sons and daughters. When Serug had lived 30 years, he fathered Nahor. And Serug lived after he fathered Nahor 200 years and had other sons and daughters. When Nahor had lived 29 years, he fathered Terah. And Nahor lived after he fathered Terah 119 years and had other sons and daughters. When Terah had lived 70 years, he fathered Abram, Nahor, and Haran. (Genesis 11:12–26 ESV)

Nevertheless, Terah is commonly argued to have been 130 years old when Abraham was born, not seventy, as this text would seem to suggest. This comes from the combination of three other passages. Firstly, the text indicates Terah died at 205 years of age:

The days of Terah were 205 years, and Terah died in Haran. (Genesis 11:32 ESV)

Shortly afterwards, the passage indicates that Abraham was 75 when he left Haran.

[4] So Abram went, as the LORD had told him, and Lot went with him. Abram was seventy-five years old when he departed from Haran. (Genesis 12:4 ESV)

At his trial, Stephen says that Abraham left Haran after Terah died:

Then he went out from the land of the Chaldeans and lived in Haran. And after his father died, God removed him from there into this land in which you are now living. (Acts 7:4 ESV)

If Terah was seventy when Abraham was born, he would only be 145 when Abraham left Haran. If he died at the age of 205, this would mean that he lived in Haran for sixty years after Abraham left. But this contradicts Stephen's claim that Abraham left after his father died.

The most common resolution to this is to argue that Terah was seventy not when Abraham was born but when an older brother of Abraham, either Nahor or Haran, was born. All three brothers were listed in the genealogy together, but presumably were not born at the same time. In order to resolve the contradiction, it is argued that Abraham was born sixty years later.

However, an alternative resolution is to argue that our copy of the text of Genesis has a scribal error. The Samaritan copy of Genesis indicates that Terah was 145 when he died, not 205. This raises the possibility that the number preserved in our texts is not the original. If so, this also resolves the contradiction and helps explain why Abraham considered himself too old to have a child.

As such, it is a matter of dispute whether or not Abraham was much older than his father when he was born. Even if he were, it would seem that that alone would give us a poor exegetical basis for concluding that the long lifespans were something other than they appear to be.

Furthermore, there is no obvious alternative interpretation of long ages. Some suggest that the ages are symbolic, are actually months rather than years, or correspond to a group of people rather than individuals. However, there is no textual indication that any of these alternative interpretations are intended.

There is an additional complication. The Greek translation of the Old Testament does not give the same ages as found in the Hebrew text. Some of the differences can be readily understood as random copying errors. However, there is a pattern; in most cases, the Septuagint indicates that the son was born one hundred years later than does the Masoretic text.

For example, while the Hebrew text indicates that Seth was born when Adam was 130, the Septuagint says that he was 230. Before the flood, the Septuagint deducts this extra century from the remaining life of the father, giving him the same lifespan. However, for the individuals after the flood, this is not deducted, recording those fathers as living for an additional century.

This systematic pattern suggests that someone deliberately changed the text to add or remove one hundred years from each of the father's ages. The natural assumption would be that the Hebrew Masoretic text was the original. However, it is not quite that simple. The Septuagint was translated from older, possibly better manuscripts, and thus may preserve the original numbers. Regardless of which numbers are original, it is difficult to see how the changes could have been accidental or who would have had the motive to make these systematic changes. As for our current subject, the great ages remain regardless of which version is taken as original.

Historical Christians

Lactantius

Lactantius discusses the long ages recorded before the Flood.[1] He cites Varro, a prolific Roman author whose works have been mostly lost. According to Lactantius, Varro sought to explain reports of ancient men living for almost a thousand years by arguing these were months and not years. Lactantius terms this explanation "manifestly false." He points out that people in the present day frequently live to be 100. This would correspond to 1200 months, but nobody is reported to have lived for

---

[1] Lactantius, "The Divine Institutes," Book II, Chapter XIII.

more than 1000 years. The ages recorded in these passages are thus not long enough to be taken as months.

Augustine of Hippo

Augustine discusses the long lifespans.[2] He invokes the great ages as part of an explanation for how Cain was able to have populated a city. But he acknowledges that a skeptic might take exception to the advanced ages.

Augustine argues that it was plausible that men used to live longer by arguing that men's bodies used to be larger. He appeals to Virgil, who in the *Aeneid* tells of a man who was easily able to lift and throw a stone that it would have taken twelve stronger modern men to lift. Augustine argues that if they were stronger men at the time of the story of the *Aeneid*, roughly 1200 BC, how much more before the flood? He also appeals to Pliny the Younger and Homer as indicating that over time, humans have experienced a decline in size.

Augustine further appeals to the experience of large bones' being found. He reports having himself seen a tooth so large that a hundred teeth could have been made out of it. He believed that the tooth belonged to a deceased giant man. While he believes the whole population of the earth was larger in the past, he believes that this giant was larger still. Today, most think that these reports of large human bones are misidentifications of the bones of large animals such as elephants.

He notes that he cannot likewise prove the long lives of the antediluvians. He insists that we must believe what the text says because the accuracy of sacred history is demonstrated by its successful predictions. He appeals to Pliny, who reported that there was a nation where men lived for two hundred years.

---

[2] Augustine, "The City of God," Book XV, Chapter 9.

If so, why should we not believe in the long lives of people who lived long ago?

Later, he discusses those who suggested that each year reported in the text is actually one-tenth of a year. He says that those who say this are "by no means to be listened to." He acknowledges that those advocating this theory are seeking to facilitate faith by making the text more believable. Nevertheless, Augustine argues that the explanation is clearly false.

In the Hebrew version of the text, taking each year to be a tenth of a year makes many much too young to be fathers. Kenan is reported to have had his son at seventy years of age. But under this interpretation, he would have his son at 7, far too young. Indeed, many are reported to have been less than one hundred years old when they fathered a son, which would make them less than 10.

This difficulty can be resolved by following the Septuagint, which adds one hundred years to many of the ages of the fathers before the birth of their son. This would make them all ten years older and thus plausibly able to father children.

Augustine responds by arguing that the Hebrew text is preferable. It is not plausible that the scattered Jews cooperated to change their own text. Rather, it is much more plausible that somebody modified an early copy of the Septuagint. Augustine suggests that it was done specifically to make this account of a ten-part year plausible by adding adding extra years to those whose ages would be too young. This explains why the Septuagint and Hebrew texts agree about Jared, because he was recorded as being 162 years of age when his son was born and thus did not need the extra years to be added.

Augustine points out that Genesis mentions months in the context of the flood narrative. In particular, it mentions the second month of the six-hundredth year of Noah's life. If the years were a tenth of a year, and thus around 36 days, what sort of months could these be? Clearly, this interpretation does not make sense.

Martin Luther

Luther wrote of the longevity of our first fathers,[3] saying, "Such longevity, when compared with the length of our lives, seems quite incredible. A question naturally arises as to the cause and theory of such old age." He favorably mentions theories that had been put forward.

He discusses the idea that they had sounder bodies than we do. He writes that the vigor and strength which man had before the fall is gone. However, their well-being was enhanced by faith in the promised seed. We have lost the strength and vigor of the first men because we have departed from their righteousness. He also speaks of their food. He suggests that a single apple in that time provided better nourishment than a thousand in our own time. The roots that they ate possessed "infinitely more fragrance, virtue, and savor" than our modern roots. Luther further extols moderation in the consumption of food. He suggests that our first parents lived moderately and ate only the food necessary to "nourish and refresh their bodies."

Luther suggests that these gifts, enjoyed in moderation, provided for long lives until God made a new order, reducing the length of man's life. However, he also says that one reason is sufficient: it pleased God to give them long lives.

John Calvin

Calvin mentions the long lifespans.[4] He makes the point that multiple generations would have known and heard from

---

[3] Luther, *Luther On The Creation: A Critical and Devotional Commentry on Genesis*, 1:Book II, Chapter II.

[4] Calvin, *Calvin's Commentary on the Bible*.

Adam directly. He simply takes it for granted that these men did live for such long periods of time.

Conclusions

Historical Christians

The historical Christians are in agreement: those living before the flood really did live for centuries. They acknowledge the surprising nature of this but reject alternative understandings. The only hint we get that at least some considered alternative readings lies in the fact that Augustine took the time to argue against the theory that each year was actually a tenth of a year.

Modern Science

The oldest person whose age is generally accepted was Jeanne Calment (1875–1997), who lived to be 122 years old. Others may have been slightly older but reliable documentation of their age is unavailable. Humans, as we know them, do not appear to be capable of living for centuries. But this does not directly tell us about the plausibility of humans' being different in some way in the past, allowing them to live for centuries.

Humans are long-lived compared to other animals. We are the longest-living land mammal. We live considerably longer than the great apes which, according to evolutionary theory, are our closest living relatives. There are other animals, such as whales, sharks, and tortoises, that live longer than humans, but even they do not attain the length of life attributed to the early humans. There are still other animals, like the hydra, that do not appear to age at all.

From an evolutionary perspective, where humans evolved from the great apes, there is little room to claim humans have lived for centuries. Instead, we would expect humans to have lifetimes in line with the rest of the animal kingdom and probably similar to those of the great apes. Humans may have evolved longer lifespans over time, but a former golden age of long lifespans is implausible.

Evaluation of the Response

The claim that antediluvians lived for centuries is as controversial today as it was in the classical world. Then, as now, it went against human experience to claim that people once lived for centuries. It is controversial within contemporary Christian circles how to understand the Biblical text. Some accept that people really did live for centuries and seek to explain this by differences in climate or by arguing that the human race has degraded, such as by the accumulation of mutations in the genome. Others argue for some other interpretation of the text.

Historically, we see that Christians accepted the authority of the text. No Christian suggested that the Biblical text was incorrect. It appears that some attempted to explain the long ages by arguing that they were referring to months; however, the authors we have looked at rejected any such interpretation.

Chapter 26

Did Methuselah Survive the Flood?

The Issue

Methuselah is well known as the Biblical character with the longest lifespan, living for nearly a millennium at 969 years. His son, Lamech, was born when he was 187. Lamech had his son, Noah, when he was 182. The flood came in the 600th year of Noah's life. If we add these numbers up we find out that the flood happened 969 years after Methuselah was born. Given that Methuselah also lived for 969 years, this suggests that he perished in the flood or shortly before it.

However, the Septuagint gives somewhat different numbers. It records that Methuselah was 167 when he fathered Lamech, twenty years younger than the Hebrew text. Lamech, on the other hand, was 188, six years older, when he fathered Noah. Together, these two differences make the timing of the flood fourteen years earlier, at 955 years after the birth of Methuselah, but this would imply that Methuselah lived for fourteen years after the flood.

The flood account does not include Methuselah among those on the ark:

> On the very same day Noah and his sons, Shem and Ham and Japheth, and Noah's wife and the three wives of his sons with them entered the ark, (Genesis 7:13 ESV)

Peter explicitly states that there were eight people on the Ark:

> because they formerly did not obey, when God's patience waited in the days of Noah, while the ark was being prepared, in which a few, that is, eight persons, were brought safely through water. (1 Peter 3:20 ESV)

Eight people would correspond to Noah, Ham, Shem, and Japeth, as well as their four wives. There does not appear to be any textual room for thinking that Methuselah was on the ark.

The simple explanation is that the Septuagint contains transcription errors. At some point in time, the numbers in the genealogies got miscopied, leading to this discrepancy. This might have happened either in the texts that the Septuagint was translated from or in early copies of the Septuagint. In fact, not all copies of the Septuagint agreed, and some texts, albeit a minority, had numbers that matched the Hebrew text.

However, this explanation was not so readily accepted in the early Church. The early Church treated the Septuagint as Scripture and many believed it had been divinely translated. When differences were found between the Hebrew and Greek genealogies, many early Christians accused the Jews of corrupting their own texts. They insisted in many cases that the Septuagint was correct. However, in this particular case, that conclusion is difficult to justify because Methuselah's lifespan does not add up.

An alternative explanation would be that Methuselah survived the flood but not on board the ark. This seems far-fetched and implausible, but it is at last conceivable that God took Methuselah to some other place to avoid the flood waters.

Historical Christians

Eusebius of Caesarea

Eusebius compares the genealogies of Genesis found in the Hebrew, Septuagint, and Samaritan texts.[1] We no longer have the original Greek text, but it survives in an Armenian translation. Our copy of his text appears to have numerical transcription issues of its own, as the numbers in his discussion do not make sense. He appeals to other copies of the text that have Methuselah living for twenty years less, making up for Lamech's being born twenty years earlier. When discussing the Hebrew version, he notes that Methuselah died at the time of the flood. However, he argues that the Septuagint is still more reliable in its record of the genealogies.

Hilary of Poitiers

Hilary briefly touches on this issue while discussing the word *homoousios* from the Nicene Creed.[2] He argues that we ought not to reject the word simply because some took an erroneous interpretation of it. He gives a lengthy list of Biblical passages which might be misunderstood. He asks if we should throw out Genesis, "lest the years of Methuselah extend later than the date of the deluge." Unfortunately, he does not elaborate on how we ought to understand this statement. However, in context, this is implied to be a false statement and thus Hilary is implying that Methuselah did not live past the flood.

---

[1] Eusebius, *Chronicle*.

[2] Hilary of Poitiers, "On the Councils," Section 85.

Ambrose of Milan

Ambrose briefly comments on Methuselah,[3] saying that his lifespan extended beyond the flood. Ambrose states that Methuselah did not feel the deluge. His mention is very brief, suggesting that he assumed his audience was already familiar with Methuselah's having outlived the flood.

Augustine of Hippo

Augustine discusses this question in several places.[4] He describes it as a frequently asked question and one on which Christians could disagree. He mentions that some believed that Methuselah was taken to be with his father, Enoch, who was taken to heaven without dying. He states that anyone may accept or reject this idea as they see fit. However, Augustine favored the conclusion that the Hebrew text and the minority of Greek texts that agreed with it were correct. Methuselah died before or in the flood.

Jerome of Stridon

Jerome discusses the issue.[5] He notes that in the Septuagint, Methuselah survives the flood by fourteen years. However, in the Hebrew text, he dies the time of the flood. He concluded that the Septuagint numbers were erroneous. Jerome made the Latin translation called the Vulgate and used the Hebrew numbers. The Vulgate would become the standard Bible of the Western Church until the Reformation.

---

[3] Ambrose, *Commentary of Saint Ambrose on the Gospel According to Saint Luke*, Commentary on Luke 3.

[4] Augustine, "On The Grace of Christ, and on Original Sin," Book II, Chapter 27; Augustine, "Questions on Heptatechum"; Augustine, "The City of God," Book XV, Chapter 11; Augustine, Book XV, Chapter 13.

[5] Jerome and Hayward, *Saint Jerome's Hebrew Questions on Genesis*.

Bede

Bede discusses the issue.[6] He mentions the differences between the Septuagint and the Hebrew and then quotes Augustine's discussion.

Summary

Historical Christians

Eusebius, Augustine, Jerome, and Bede all appeal to a textual copying error to resolve the question. They differ in whether or not they take the Septuagint to be in error or favor the minority of texts without this issue. Hilary implies that we are not to understand Methuselah as surviving the flood. Ambrose thinks that Methuselah did survive the flood.

Evaluation of the Response

Historical Christians took seriously the apparent Biblical claim that Methuselah survived the flood. They were unwilling to accept that the text was simply incorrect. In this, they demonstrated an acceptance of the authority of the Biblical text.

The apparent problem was resolved by going back to the source. The problem was caused by transcription errors in the text. The problem is easily resolved by returning to the original language text, in which Methuselah perished about the same time as the flood, rather than surviving it.

This should also encourage us. Historical Christians were able to resolve this problem by careful study of the Biblical text. They did not have to abandon the authority of the text or adopt a doubtful interpretation of the text.

---

[6] Bede, "Commentary on Genesis."

Chapter 27

Was the Ark Big Enough?

The Biblical Question

The story of Noah's flood is one of the best-known Bible stories. According to it, two of every animal were preserved on the Ark. How big was the Ark?

> This is how you are to make it: the length of the ark 300 cubits, its breadth 50 cubits, and its height 30 cubits. (Genesis 6:15 ESV)

A cubit was the length of a forearm, approximately one and a half feet or half a meter. This makes the Ark 450 feet (150 meters) long, 75 feet (25 meters) wide, and 45 feet (15 meters) tall. This is somewhat larger than the largest wooden boat constructed in modern times. However, it would be much smaller than a modern cruise ship.

The Ark was quite large for a boat. However, there are numerous different kinds of animals. Was there enough room on the Ark for all of them? What about the very large animals? What about the food required to sustain the animals?

Historical Christians

Origen

Origen engages with heretics who sought to disparage the Old Testament by arguing that the flood account was obviously false because there was no way all of the animals could fit.[1] He also responds to Celsus, who objected that the Ark was absurdly large.[2] Origen says that Celsus got the objection wrong; he should have objected that the Ark was not large enough to house all of the animals.

Origen argues that Moses, being trained in all the "wisdom of the Egyptians," used an unusual method of numerical calculation. He argues that when Moses said 300 cubits, this actually meant three hundred squared or 90,000 cubits. This would mean the Ark was not 450 feet (150 meters) long but 25 miles (40 km) long. This would make the Ark large enough to be a city.

Augustine of Hippo

Augustine discusses this issue.[3] He mentions those who object that the Ark was not large enough to contain so many animals. He suggests that they neglect to consider that the Ark was made of multiple floors. He also mentions possibly accepting Origen's interpretation. However, Augustine was using a Latin translation of Origen, which indicated that each cubit was six cubits, giving much smaller numbers than Origen suggested. Augustine also insists that it would be possible to build such a large Ark. He points out that large cities have been built, and the Ark is stated to have taken over one hundred years to build.

---

[1] Origen, *Homilies on Genesis and Exodus*, Homily II.

[2] Origen, "Contra Celsum," Book IV, Chapter 41.

[3] Augustine, "The City of God," Book XV, Chapter 27.

John Calvin

Calvin discussed this issue in his commentary on the passage.[4] He says that "the question respecting its magnitude is more difficult." He says that some have ridiculed such a multitude of animals shut up in such a small place. He notes the idea from Augustine and the translation of Origen which indicated that each cubit was six cubits. However, he rejects this because Moses always wrote in a plain style to be understood and did not use strange systems of numbers. Calvin indicates that he does not know the size of the cubit used. He concludes that God knew the appropriate size for the Ark and that the story is filled with miracles.

Conclusions

Historical Christians

All of the authors considered assumed that the flood narrative took place and that the surviving animals were all on the Ark. They seek to answer the question of how so many different animals fit on the Ark. Origen appeals to a colossal Ark. Augustine appeals to the idea that the Ark may have had multiple floors but also makes room for the notion that the cubits mentioned in the text may have been larger than the cubits of a later time. Calvin rejects implausible alternative cubits, but does not provide a different interpretation of his own other than to suggest the text is full of miracles.

Modern Science

We have discovered many more kinds of animals than would have been known to historical Christians. For example, none

---

[4] Calvin, *Calvin's Commentary on the Bible*.

of the authors would have known about llamas. Furthermore, they were not aware of the wide variety of different species of familiar animals that existed. There are, for example, hundreds of thousands of species of beetles. The sheer number of different animals that have to be accounted for is extremely large.

However, modern science has also developed a theory of evolution. Biological species do change over time. Evolution as the complete explanation for the origin of all species is controversial among Christians. However, most Christians would accept micro-evolution or the small-scale development of new species from older species. This acts to reduce the number of species that would have been on the Ark, as many of the species we know could have evolved from similar species.

Evaluation of the Response

The question of whether or not the Ark would be able to house all of the animals is controversial today. Modern young-earth creationists have argued that there was indeed sufficient room as long as we accept that not every individual species had to have been on the Ark.[5] Critics have argued that this is simply not possible.[6]

Historical Christians accepted the Bible's authority. The Bible indicated that the Ark carried all of the animals, and thus they insisted that it did. Even in the face of criticism, they insisted on the truth of the Biblical account. None of the authors suggested that the Bible was incorrect. None suggested that the passage was solely allegorical and did not describe a historical event.

---

[5] Woodmorappe, *Noah's Ark*.
[6] Moore, "The Impossible Voyage of Noah's Ark."

Chapter 28

Are Moneylenders Parasites?

The Scientific Question

Moneylending has been present throughout recorded history. But what does this profession actually contribute to society? The shepherd clearly contributes by producing food and wool. The tailor contributes by producing clothes. The soldier contributes by defending his homeland. In contrast, the moneylender does not appear to directly produce any useful goods or services to society.

Is the moneylender then simply a parasite on society? Is the moneylender simply taking advantage of his fellow man, somehow exploiting them for his own gain? If not, is the moneylender somehow contributing to society in an indirect way? Does the moneylender's choice about who to loan money to actually help and contribute to society?

Crucial to the question is the charging of interest, also known as the practice of usury. Moneylenders do not expect to simply be paid back the same amount that they lent out. Rather, they expect to be paid back that amount as well as an additional amount known as interest. The interest is essentially the price of the loan.

Lending money at interest was controversial in the ancient world, with influential philosophers such as Cato the Elder, Cicero, Seneca the Younger, Plato, and Aristotle all condemning it. We are not concerned here with the morality of charging interest. That is not a scientific question. However, the morality of charging interest is bound up in the question of whether the moneylender is performing a useful service. Generally, the condemnation of charging interest is rooted in the belief that moneylending is not helpful to society but actually a detriment to it.

In some cases, loans are made from wealthy individuals to poor individuals who are in a desperate situation. They may have sudden medical expenses, have to deal with a failed crop, or experience any number of other disasters. Whatever the situation, they have expenses they cannot pay for and go to a moneylender for a loan to cover those expenses. But the end result is that they have to pay that loan back later, plus interest. The end result is often that the poor get poorer and the rich get richer.

However, this is not the only type of situation in which loans occurred in the ancient world. Another common kind of loan was when a merchant borrowed money in order to fund a maritime trade expedition. They would take out a loan in order to purchase goods in one place and transport them via ship to another place, where they would sell them. They would then repay the moneylender, and both the moneylender and the merchant would profit. The availability of a loan enabled a profitable venture and thus both the lender and merchant benefited from the transaction.

Aristotle argued that charging interest was unnatural. In his view, the purpose of money was to be used as a medium of exchange. One household would sell what it had too much of for money in order to buy what it needed. However, he deemed using money beyond that to be unnatural. Charging interest was the most unnatural use of money of all.

The Biblical Discussion

The Old Testament contains condemnations of those who lend out their money at interest and praise for those who do not:

> who does not put out his money at interest and does not take a bribe against the innocent. He who does these things shall never be moved. (Psalm 15:5 ESV)

> In you they take bribes to shed blood; you take interest and profit and make gain of your neighbors by extortion; but me you have forgotten, declares the Lord GOD. (Ezekiel 22:12 ESV)

Ezekiel goes so far as to condemn the taking of interest right along with taking bribes to shed blood. It would be very easy to conclude from these verses that taking interest is simply immoral. However, such a simplistic conclusion cannot be maintained because Deuteronomy endorses charging interest, even if only to foreigners:

> You shall not charge interest on loans to your brother, interest on money, interest on food, interest on anything that is lent for interest. You may charge a foreigner interest, but you may not charge your brother interest, that the LORD your God may bless you in all that you undertake in the land that you are entering to take possession of it. (Deuteronomy 23:19–20 ESV)

If the charging of interest were simply immoral, then it would still be immoral to charge interest to foreigners. The two earliest prohibitions on charging interest in the Law specifically forbade loans to poor Israelites:

> If you lend money to any of my people with you who is poor, you shall not be like a moneylender to him, and you shall not exact interest from him. (Exodus 22:25 ESV)

> If your brother becomes poor and cannot maintain himself with you, you shall support him as though he were a stranger and a sojourner, and he shall live with you. Take no interest from him or profit, but fear your God, that your brother may live beside you. You shall not lend him your money at interest, nor give him your food for profit. (Leviticus 25:35–37 ESV)

Possibly the later condemnations of interest should be read in light of the qualifications in this early prohibition. If the concern is specifically lending to poor brethren, taking advantage of their desperate situation, then other loans, such as for commercial enterprises, would be allowed. This may explain the exception for foreigners, who would likely be merchants interacting with Israel.

The only New Testament references to interest are found in a couple of Jesus' parables.

> But his master answered him, 'You wicked and slothful servant! You knew that I reap where I have not sown and gather where I scattered no seed? Then you ought to have invested my money with the bankers, and at my coming I should have received what was my own with interest.' (Matthew 25:26–27 ESV)

> He said to him, 'I will condemn you with your own words, you wicked servant! You knew that I was a severe man, taking what I did not deposit and reaping what I did not sow? Why then did you not put my money in the bank, and at my coming I might have collected it with interest?' (Luke 19:22–23 ESV)

A simple reading suggests that Jesus is endorsing the lending of money at interest. However, the master is explicitly condemning the servant with the servant's own words. He is stating what the servant should have done if the servant really believed his accusation that the master sowed where he did not reap. As such, it is not clear that the master or Jesus were actually endorsing lending out money at interest.

However, neither parable actually mentions loans. The text does mention bankers who pay interest on money deposited with them. However, bankers at this time did not primarily make money through loans. Rather, their primary business was money-changing, exchanging one currency for another. They did take deposits and pay interest, but primarily to fund money-changing and not loans. The Greek word translated as bank is τράπεζα (trapeza), and it literally means table. This is because they did their business on tables. We read about these tables elsewhere:

> And Jesus entered the temple and drove out all who sold and bought in the temple, and he overturned the tables of the money-changers and the seats of those who sold pigeons. (Matthew 21:12 ESV)

The bankers whom Jesus refers to are probably money-changers and not moneylenders. Jesus does condemn the money-changers when he chases them out of the temple. However, the only clear indication from the text of the cleansings of the temple is that Jesus objects to money-changing's being done in the temple. It is not clear that Jesus objected to the money-changers' trade in principle.

But the Biblical injunctions actually go beyond simply not lending at interest. Every seven years, there was a release of all debts:

> At the end of every seven years you shall grant a release. And this is the manner of the release: every creditor shall release what he has lent to his neighbor. He shall not exact it of his neighbor, his brother, because the LORD's release has been proclaimed. (Deuteronomy 15:1–2 ESV)

The implication is that the lender would not only not receive interest; he would not get back the loan itself. However, the text goes on to insist that the Israelites are obligated to lend to their brothers even if the year of release is near:

> If among you, one of your brothers should become poor, in any of your towns within your land that the LORD your God is giving you, you shall not harden your heart or shut your hand against your poor brother, but you shall open your hand to him and lend him sufficient for his need, whatever it may be. Take care lest there be an unworthy thought in your heart and you say, 'The seventh year, the year of release is near,' and your eye look grudgingly on your poor brother, and you give him nothing, and he cry to the LORD against you, and you be guilty of sin. (Deuteronomy 15:7–9 ESV)

Jesus makes a similar statement:

> And if you lend to those from whom you expect to receive, what credit is that to you? Even sinners lend to sinners, to get back the same amount. But love your enemies, and do good, and lend, expecting nothing in return, and your reward will be great, and you will be sons of the Most High, for he is kind to the ungrateful and the evil. (Luke 6:34–35 ESV)

Jesus instructs us to lend and expect nothing in return. We must be willing to help others out, even at great personal cost.

The Biblical principle is that we should not look at someone in need as an opportunity to make money off of their misfortune. Instead, we should be seeking to help them out to our detriment. The prohibition on charging interest should be understood as part of this principle. It does not reflect a belief that charging interest is unnatural or innately immoral. Rather, it recognizes that loaning money at interest is, in some cases, taking advantage of someone in a desperate situation.

Historical Christians

Leo the Great

Leo the Great condemns the moneylender's trade.[1] He declares that it is a sin "to either lessen or increase the sum" and declares that he who takes more than he lent is wretched. Usury of money, he declares, is the death of the soul. He cites Psalm 15 which speaks of those who lend money out at interest not being allowed to rest on God's holy hill. He describes usury as deceitful gain and as enriching oneself by other people's losses. As a result, those who engage in usury will be punished with eternal neediness.

Augustine

Augustine comments on usury.[2] In a modern English translation, the verse in question says:

> ruin is in its midst; oppression and fraud do not depart from its marketplace. (Psalm 55:11 ESV)

However, the Septuagint translated oppression as usury. As such, this passage does not actually speak of usury in the original Hebrew, although other passages do. Augustine laments that usury is not only practiced but done so openly. He goes on to describe an even worse form of usury: failing to forgive others. He cites the Lord's prayer, "forgive us our debts as we

---

[1] Leo the Great, "Sermons of St Leo The Great," Sermon 17.
[2] Augustine, "Exposition on Psalm 55."

too forgive our debtors." He insists that we ought to "consent to exact just so much as you have given." For Augustine, failing to forgive someone is demanding more than one is owed.

Thomas Aquinas

Aquinas discussed the morality of usury.[3] He cites Exodus's prohibition on usury to show that it is a sin to take interest. He argues that charging interest is unjust because it sells what does not exist. Some things, like food or wine, are consumed during their use. It makes sense to charge for them, but it would not make sense to charge for their use. It would be nonsensical to rent food. However, other things, like houses, are not consumed in their use. In this case, it makes sense to charge for using the thing. However, money was meant to be used to purchase things. Aquinas cites Aristotle, arguing that money is intended to be a medium of exchange. But this means that money is consumed in its use and thus it is like food and not houses. As such, it is unnatural to charge for its use.

Aquinas considers an objection based on the parable where the master recommends depositing money with the bankers. If Jesus condones earning interest on money, it cannot be a sin. Aquinas answers that the interest in the passage should be taken figuratively as a reference to an increase of spiritual goods.

Another objection points out that charging interest was allowed to Gentiles and that another verse actually promises that Israel would lend to many nations if they followed his law.

> The LORD will open to you his good treasury, the heavens,
> to give the rain to your land in its season and to bless all the

---

[3] Aquinas, *The Summa Theologiæ of St. Thomas Aquinas,* The Second Part of the Second Part, Question 78.

> work of your hands. And you shall lend to many nations, but
> you shall not borrow. (Deuteronomy 28:12 ESV)

The Vulgate translates "lend" using a word that specifically refers to lending at interest. If God allowed lending at interest to other nations and even promised that Israel would lend at interest, how can it be wrong to take interest? Aquinas insists that it was still wrong to charge interest to Gentiles, but it was allowed by the law in order to avoid the greater evil of the Israelites' taking interest from their fellow countrymen. As for the promise that Israel would lend to other nations, Aquinas argues that it refers to lending without charging interest because Israel would be so wealthy.

Another objection points out that civil laws do not forbid usury. Aquinas argues that the civil law leaves some things unpunished that are immoral. He says that those who are imperfect would be deprived of many advantages if all sins were strictly forbidden and punished by civil law.

Another objection cites Jesus' instruction to lend and expect nothing in return, contending that this passage is advice rather than a command. As such, it is not sinful to not follow this passage. Therefore, Jesus is advising against lending at interest, not condemning it as immoral. Aquinas responds that it is advice because no one is obligated to lend money at all. Nevertheless, it is immoral to seek profit by lending.

Another objection argues that it cannot be sinful to accept a price for what one is not bound to do. Since we are not bound to lend to our fellow man, it cannot be wrong to accept a price for lending it. Aquinas argues that repaying the loan without interest is the appropriate price of the loan.

Another objection argues that it would be moral to rent out a silver vessel. As such, it should also be moral if the same silver were made into coins. Aquinas argues that these cases are different because the vessel is not used up while it is used. However, money is used up by being spent as it is used.

Another objection argues that because the lender willingly accepts the terms of the loan, it cannot be that it is immoral. Aquinas responds that the borrower's acceptance of the terms are not strictly voluntary because he needs to borrow money and the owner is unwilling to lend without charging usury.

Martin Luther

Luther wrote a treatise attacking usury.[4] He decries the fact that usury is everywhere while its purveyors purport not to be engaged in usury. He defines three degrees of dealing with worldly goods. The first degree requires us not to resist those who would take from us, as expressed by Jesus:

> And if anyone would sue you and take your tunic, let him have your cloak as well. (Matthew 5:40 ESV)

The second degree requires that we give to all who need it. The third degree is that we lend willingly and gladly without charges or interest. He supports this by appealing to Scripture:

> Give to the one who begs from you, and do not refuse the one who would borrow from you. (Matthew 5:42 ESV)

> If among you, one of your brothers should become poor, in any of your towns within your land that the LORD your God is giving you, you shall not harden your heart or shut your hand against your poor brother, but you shall open your hand to him and lend him sufficient for his need, whatever it may be. (Deuteronomy 15:7–8 ESV)

Luther points out that the historical Christian theologians had all agreed in condemning usury, although they had not agreed about whom we were obligated to lend to. Luther

---

[4] Luther and Jacobs, "A Treatise on Usury."

cites Jesus' command to lend expecting nothing in return. He argues that we are obligated to lend even to our enemies. He acknowledges that some have taken this passage as forbidding lending at interest, but while he believes that charging interest is prohibited, he argues that if one charges interest, this is not loaning but usury. As such, when Jesus speaks of loaning, he, by definition, is not talking about usury. He goes on to reject any suggestion that this should be taken as advice rather than a command.

Luther appeals to the Golden Rule, "Do unto others what you would have them do unto you." Nobody would want to pay back more than they borrowed. Therefore, demanding more in return than was loaned is against nature and a mortal sin. He also appeals to the command to love one's neighbor as oneself. These lenders, Luther insists, love only themselves.

Luther considers the objection that if money were not lent at interest, those lending the money would lose because of opportunity cost. Any individual with money could, instead of lending their money, invest it in some form of business and make money on it. As such, by lending the money instead, they lose relative to what they could have made. As such, it is only fair that they be compensated for this lost profit. In response, Luther points out that the would-be lender would lose both the money and potential profit if it were stolen from them or they freely gave it away. If the profit is lost in those cases, on what basis do they think they are owed profit when money is lent out?

Luther goes on to consider the practice of buying an income, that is, purchasing an investment that pays a set return over time. Luther sees this as immoral but the immorality is hidden. Firstly, he objects that it is a "new and slippery invention, especially in these last, perilous times, where nothing good is invented anymore and the thoughts of all men are bent upon wealth and honor and luxury, without any limit." Secondly, he objects that it has a negative appearance and we are commanded to avoid all evil appearances. He is referring to:

Abstain from every form of evil. (1 Thessalonians 5:22 ESV)

However, some will be familiar with other translations that instruct us to avoid every appearance of evil. Nevertheless, almost all modern translations translate the word as form or kind. The verse is probably instructing us to avoid all kinds of evil and not merely the appearance of evil.

Luther points out that one always prefers to be the buyer rather then seller of such incomes. This demonstrates that it is done for the benefit of the buyer. He says that everyone's conscience fears the morality of buying income, while nobody fears selling income. Luther acknowledges that the business can be done without charging usury. However, he insists that it cannot be done with appropriate love and care for the neighbor. Nobody would choose to be the seller, having to provide an income, instead of the buyer receiving an income. He states that "it lays burdens upon all lands, cities, lords, and people, sucks them dry and brings them to ruin." If the fruit of the selling of incomes is so bad, then the root must be bad as well.

Luther considers the argument that given one hundred gulden, a coin, it would be possible to use it in trade and labor to produce five or six guldens. When someone buys an income with the one hundred gulden, he is simply collecting the income that he could have earned using those coins. Luther says, "All that is so pretty that no one can find fault with it at any point."

Luther points out that engaging in any sort of business brings risk. The one engaging in business might well lose the money instead of making additional money. He calls this counter-interest. Furthermore, he suggests that the counter-interest corresponding to this risk ought to be considered greater than the interest gained through productive use of the funds.

John Calvin

In his commentary on Exodus 25:35,[5] Calvin wrote that "Humanity ought to be very greatly regarded" with respect to loans, especially when someone in an extreme situation depends on the compassion of a wealthy man. Calvin writes that this is the true test of our charity, when, according to Christ's command, we lend to those from whom we expect no return. However, Calvin writes that the question here is not about usury. Christ is commanding us not to loan only to the rich, from whom we might expect a return, but also to the poor, from whom we expect no return and may not even receive our money back.

Calvin points out that the command about usury follows the previous sentence, which commands supporting fellow Israelites who are poor. The rich are commanded to help uplift the poor. The extraction of interest will "entirely drain" the poor. God's object in giving this law was to encourage brotherly affection among the Israelites. He notes that Israel was allowed to lend to Gentiles, but the distinction between Jew and Gentile has been abolished in the new covenant, and thus we ought not to grow rich by the loss of others. He writes that the Jews were allowed to charge interest to Gentiles because it would be unfair if the Jews were not allowed to charge interest to Gentiles while Gentiles were able to charge interest to Jews.

Calvin then asks whether usury is evil in and of itself. He points out that it was widely condemned by the ancient pagans. Nevertheless, our judgment must be based on the

---

[5] Calvin, *Calvin's Commentary on the Bible.*

universal law of justice, "Do unto others as you have them do unto you." It does not matter what we call usury, and crafty attempts to call usury by another name do not change the situation. We have to consider the facts of the matter, not the names that are given to it. He points to Ezekiel 18:13 which condemns usury and taking profit. The word translated as usury is נֶשֶׁךְ (neshek) which derives from a word meaning to bite. Calvin argues that because of the negative connotations of this word, they called the interest by another word, תַּרְבִּית (tarbith) which is derived from a word meaning to increase. However, Ezekiel condemns both to make it clear that it does not matter what we call it.

However, Calvin argues that reason does not permit us to think that usury is be condemned without exception. If the debtor causes his creditor loss and inconvenience, then it only makes sense that the creditor should be compensated for his loss. Further, if a rich man borrows to buy a piece of land, it is only reasonable that the lender should receive part of the revenue gained from the land.

Calvin discusses Aristotle's argument that money is barren and that it is thus unnatural to earn interest on money. He points out that much profit can be had by trading with another man's money or purchasing a farm. Essentially, money is not barren and thus Aristotle's argument is based on a false premise.

However, others will object that God prohibits usury completely. Calvin argues that this applies only to the poor. With the rich, usury is freely permitted. With respect to passages which seem to absolutely condemn usury, Calvin says that their statements "ought to be judged by the rule of charity." Only unjust exactions are condemned, not usury of all sorts.

Nevertheless, Calvin indicates that he is unwilling to engage in usury and wishes it did not exist. Yet, he will not dare go beyond what God's words convey. He holds that the ancient Israelites were prohibited from charging usury, but this was part of their particular law and usury is now lawful, except insofar as it violates "equity and brotherly union."

Conclusions

Historical Christians

The condemnation of usury is consistent among these historical Christians. Calvin alone makes distinctions and thinks that some forms of usury may be moral, although he himself will still not engage in it. There is a general belief that the moneylenders made their money through the losses of someone else.

Modern Science

Despite the condemnation of the Church, the practice of usurious loans developed over the course of the Middle Ages. Economic development made it more and more useful to engage in the practice of loaning money at interest. This would become even more pronounced with the industrial revolution and the rise of modern capitalism. By the time of Adam Smith, who is often considered the father of modern economics, the concept of charging interest on loans was accepted as good and necessary.

Modern economic science understands moneylending institutions as playing a vital role by helping to allocate resources. They do not simply extract money from others; they help others by providing money when it is needed. Without moneylending and interest, it would not be possible to get a mortgage to buy a house or a loan to start a business.

Evaluation of the Response

Historically, Christians have condemned the charging of interest on loans. Today, very few Christians take this position. Today, usury is often defined as charging an excessive rate of interest. This is not what Christians meant historically when they condemned the practice of usury; they were referring to any charging of interest. Almost all modern Christians have accepted the charging of interest.

The historical Christians interpreted the Biblical condemnations of interest in light of the classical world's condemnation of interest. They concluded, for the most part, that charging interest was inherently unnatural and immoral. They very readily understood the Bible as condemning all usury for the same reasons.

However, more careful consideration of the Biblical text could have led them to a more nuanced understanding. Aquinas's responses to Biblical objections were not very compelling. Calvin admitted that a universal condemnation of usury was not supported by the Biblical text. A careful return to the source would have revealed a more nuanced understanding of Biblical teaching.

Chapter 29

Do Partridges Steal Eggs?

The Biblical Discussion

The partridge is a medium-sized bird similar to a pheasant or chicken. It is best known for its recurring role in the song "The Twelve Days of Christmas." They are notable for making their nests on the ground rather than in trees, and for producing a large number of eggs. In fact, some have been known to engage in double-nesting, where both the male and female brood over separate nests.

Jeremiah speaks of the partridge:

> Like the partridge that gathers a brood that she did not hatch, so is he who gets riches but not by justice; in the midst of his days they will leave him, and at his end he will be a fool. (Jeremiah 17:11 ESV)

This would seem to describe a bird which steals the hatched baby birds from other mothers. This would be an exceedingly strange behavior. The New American Standard Bible says instead that the partridge "hatches eggs which it has not laid" suggesting that it steals eggs from other birds and hatches them, a different but equally odd behavior. The New King James Version says that the partridge "broods but does not hatch" suggesting the more plausible behavior of a bird which lays and broods over many eggs, but fails to hatch many of them.

But what does the Hebrew text actually say here? The Hebrew says that the partridge gathers, דָּגַר (dagar), but does not hatch, יָלָד (yalad).

There is only one other passage in the Bible which uses the word *dagar*:

> There the owl nests and lays and hatches and gathers (*dagar*) her young in her shadow; indeed, there the hawks are gathered, each one with her mate. (Isaiah 34:15 ESV)

The owl in this passage is brooding over her young. Brooding is a practice in birds where they sit on their eggs or young to keep them warm. The limited Biblical usage makes it difficult to be certain of the meaning of the word. The Brown-Driver-Briggs lectionary defines the word as meaning "to gather together as a brood." However, it is possible that the word does not refer to gathering but simply to brooding.

The meaning of *yalad* is straightforward; it is the commonly used word for bring forth or beget. It is widely used of a mother bringing forth a child. However, it is somewhat unclear at which point a bird is considered to have brought forth. Does the bird bring forth a child when it lays the egg? Or is the child not brought forth until the egg hatches?

The uncertainties about how to understand these verbs give rise to a few possible readings:

1. The partridge broods over but does not hatch the eggs
2. The partridge gathers a brood it did not hatch.
3. The partridge broods over eggs it did not lay.

The simplest reading is arguably the first. It states, very simply, that the partridge broods over eggs which never hatch. The other readings are less straightforward. The idea of a bird that gathers a brood but did not hatch them or a bird that broods over eggs it did not lay is more complicated. This gives us some reason to think that the first reading might be the intended reading.

The text indicates that the partridge is like a man who gets riches but not by justice. But how is the partridge like this? Possibly, the intention is that the partridge gains offspring by stealing them just as the ungodly gain riches by injustice. However, it is not clear what exactly that comparison is supposed to convey. Alternatively, the point of the comparison might be that the ungodly man will lose his riches, just as the partridge loses its eggs. Thus the text goes on to say of his riches, "in the midst of his days they will leave him, and at his end he will be a fool."

The Septuagint says:

> The partridge utters her voice, she gathers eggs which she did not lay; so is a man gaining his wealth unjustly; in the midst of his days his riches shall leave him, and at his latter end he will be a fool. (Brenton Translation)

The Septuagint adds an indication that the partridge "utters her voice." The Hebrew word for partridge, קֹרֵא (qore) derives from a word meaning to call out. This is probably a reference to the harsh cry of the partridge. The translator of the Septuagint appears to have translated the word for partridge twice, once as the Greek for partridge and once more literally as a reference to crying out.

The translator also rendered the text as indicating that the partridge gathers that which she did not bring forth. The translation quoted above indicates that she gathers what she did not lay, but the Greek is more generic. While there is some question about what Hebrew intends, the Septuagint does indicate that the partridge steals the eggs or offspring of other birds.

The Vulgate uses the word *fovit* which refers to keeping warm. This is probably a reference to the partridge's keeping the eggs warm before they hatch. Thus, the Vulgate indicates that the partridge broods over eggs that will not hatch.

Taking all of these things into account, the most likely explanation is that the original author was simply referring to the partridge as brooding over eggs that would not ultimately hatch. This fits the way these words are used in other passages, is the simplest reading, and makes sense of the passage as a whole. The idea that the partridge steals the eggs or young of other birds seems to derive ultimately from a mistranslation in the Septuagint.

The Scientific Discussion

The partridge had a very poor reputation. Aristotle wrote about partridges in his *History of the Animals.* He describes the bird as having a "malign and crafty disposition." They try to draw potential predators away from their nests by pretending to be injured and limping away from the nest, only to fly off when her young have escaped. He states that because of their lecherous nature, the males of the species try to smash the eggs. The female seeks to prevent this by laying her eggs while she runs away from the male. When the female escapes, the males will fight. He also claims that the partridges can become pregnant from simply standing downwind or under the flight of the males. However, nowhere in these ideas do we find a suggestion that partridges steal the young or eggs of other birds. This does not appear to have been an idea found in the classical world.

Historical Christians

Hippolytus

Hippolytus wrote about this verse.[1] He describes the partridge as an overly vain creature. He writes that when a partridge sees nestlings unattended by their mother, it imitates the call of their mother, drawing the birds away from their nest into its own. However, when the real mother returns, the nestlings recognize her as their real parent, and they flee back to her. Hippolytus takes this as a parable of the Antichrist, who likewise will draw mankind to himself.

Origen

Origen discusses the partridge,[2] calling it a famous question. He describes the animal as wily and unscrupulous and wanting to deceive hunters. He says that the mother rolls around at the feet of the hunter to distract him so that her young may escape. He also accuses the bird of being indecent, claiming that the males fight each other for sex and engage in homosexual intercourse.

Since the partridge is a malicious, deceitful, and indecent animal, it is clear that it cannot be a symbol for Christ but must be a symbol for the Devil. He compares the partridge to the Devil who does not gather what he has generated but steals the creatures belonging to another.

Ambrose

Ambrose discusses the partridge while discussing a number of creatures that God made.[3] He says that the partridge is

---

[1] Hippolytus, "On Christ and Antichrist," Section 55.

[2] Origen, *Homilies on Jeremiah and I Kings 28*, 97:Homily 17 on Jeremiah.

[3] Ambrose of Milan, "Hexameron," Homily 9.

cunning and steals the eggs of another partridge. However, as soon as the eggs hatch, the chicks follow a natural instinct and return to their true mother. Ambrose cites Jeremiah as referring to this practice. The brooding does not benefit the partridge but the true mother.

Jerome

Jerome discusses the partridge in his commentary on Jeremiah.[4] He cites the writers of natural history, specifically mentioning Aristotle, Theophrastus, and Pliny the Elder, as recording that it is the nature of the partridge to steal the eggs of other partridges. Jerome is incorrect, as none of these authors appears to have recorded any such thing. The partridge sits on the eggs, keeping them warm, but when the eggs hatch, the chicks fly away from them. Jerome says that this is like the rich, who "get riches but not by right." They are struck by sudden death and lose their riches.

John Calvin

Calvin wrote about the partridge in his commentary on Jeremiah.[5] Calvin notes that interpreters do not agree on how to read the passage. However, the majority read the partridge as stealing the eggs of others, just as the wicked rich man accumulates wealth. However, in the end, neither the partridge nor the rich man will have anything.

Calvin does not accept this interpretation, attributing it to a fable by Jewish rabbis. He notes that neither Aristotle nor Pliny says any such thing about the partridge. He describes the idea that partridges would steal eggs as very absurd to believe. He takes the text as indicating that the partridge broods over eggs,

---

[4] Jerome, *Commentary on Jeremiah*.
[5] Calvin, *Calvin's Commentary on the Bible*.

which do not end up hatching. In particular, he takes this to be because the male partridges smash the eggs of the females so that they can fulfill their lustful urges.

Conclusions

Historical Christians

The various authors take slightly different accounts of the behavior of the partridge from this verse. Hippolytus thinks that the partridge steals the chicks. Ambrose and Jerome think that the partridge steals the eggs. The later authors attribute this idea to the writings of natural historians. Calvin rejects the idea of theft and thinks instead that the males destroy the eggs.

Modern Science

Partridges do not steal the eggs or young of other birds. Neither do any other birds. There are birds that do the opposite, sneaking their eggs into the nests of other birds. These brood parasites are thus able to have children while having other birds take care of the young. Logistically, it is difficult to see how a partridge could even transport eggs from another bird's nest to her own. Furthermore, the idea seems to make little sense strategically. Raising the offspring of another bird would be an act of altruism, not the act of a bird's trying to gain from deceit.

Evaluation of the Response

Most of the historical Christians considered were incorrect; the partridge does not steal either young chicks or eggs. Calvin alone rejects the incorrect story, although his account of males' destroying eggs is not correct either. Nevertheless, we can identify two reasons for this error. Firstly, they were getting information from the Septuagint, which incorrectly translated this verse. The Septuagint appeared to be describing a behavior which does not actually exist. Secondly, while the ancient philosophers did not report this particular behavior, they did view the partridge very negatively, predisposing these historical Christians to accept this story about the partridge.

The story illustrates, yet again, the importance of returning to the source. The idea derives from a flawed translation of the Septuagint. Failure to return the Hebrew text and evaluate its meaning leaves us with a text that appears to make inaccurate claims about the behavior of the partridge. Furthermore, they failed to go back to classical sources to verify claims that they reported the same behavior.

Chapter 30

Lessons Learned

Introduction

What have we learned from all of this history? We have looked at the nature, history, shape, movement, and firmness of the heavens. We've considered the waters above, the waters below, the waters in the ocean, the waters covering earth, four rivers of waters, and springs of water. There is whether people lived long lives before the flood, whether Methuselah survived without the ark, and how big the ark was. We've learned about the sizes of the heavenly bodies, the source of the moon's light, the effects of the stars on our actions, and the nature of the Star of Bethlehem. We've asked whether usury is unnatural and whether partridges steal the eggs of other birds. But what, in the end, have we really learned?

Take Heart

The first thing we should learn is to take heart. It can be easy to become discouraged as a Christian today, having to defend against the onslaught of scientific arguments against our faith. It is tempting to think that we are the first Christians who have had to face scientific challenges to our faith. We live in an era when the traditional Christian understanding of origins is seen as an ignorant myth, the Bible is ridiculed as filled with scientific flaws, and Christian moral claims are seen as baseless prejudice. According to a prevalent cultural narrative, the Christian faith is fighting a losing war with science, losing a series of skirmishes over issues such as heliocentrism, the age of the earth, and evolution on its way to irrelevance.

The Church itself is contentiously divided. Some advocate for reevaluating our fundamental doctrines in light of modern science, while the other extreme is willing to reject even such seemingly established scientific facts as heliocentrism and a spherical earth. Since we think of science as a fundamentally modern phenomenon, it would be easy to assume that historical Christians did not have to face such issues. But as we have seen, this is clearly not the case. There is nothing new under the sun; the Church faced just such attacks in its early years.

The Bible made a variety of claims that were at odds with classical science. The Bible indicates that the heavens are changeable, in contrast with the immutable heavens of Aristotle. Christian theology insists that God created the world *ex nihilo*, in contrast with the dictum that nothing comes from nothing. The Bible narrates the creation of man from the dust of the ground in contrast with the conception of the human race as eternal. The Scriptures refer to the waters above, in contrast with the insistence that waters could not remain above the stars. There was a widespread belief in the ancient world that the stars dictated events on earth, but Christianity rejected this idea. Classical science contradicted Biblical claims on a number of points.

A minority of authors interpreted the Bible as going much further in rejecting classical science, teaching what we now think of as flat-earth cosmology. This included an insistence that the bottom of the earth could not be the center, as conceived of by classical science. Likewise, the heavens were not a sphere rotating around the earth. The earth was alleged to float on the ocean. However, this response was a minority view and its adherents were unable to convince other Christians that the Bible taught their views.

A number of objections could be raised against details of the Biblical account of early history. The Bible claims the sun was created in the middle of Creation Week, despite the sun's seemingly being required earlier in the week. According to the Septuagint, God made birds out of water when either air or earth would have been more appropriate. Genesis records a spring that watered the whole earth and four rivers flowing out of Eden, which match no known geography. Early chapters in Genesis record people living for centuries, far longer than any recorded humans. It seemed like the Ark would not be big enough to include every kind of animal. The early history in Genesis was open to objection on the basis of classical science.

Furthermore, certain Biblical statements could be interpreted as contradicting classical science. The Biblical narration of the creation of the sun, moon, and stars could be taken as making incorrect claims about the relative sizes of astronomical objects. The description of the moon could be taken as implying that the moon generates its own light instead of reflecting the sun's light. Biblical references to the pillars of the earth could be understood as teaching that the earth rests on pillars. Christians rejected these interpretations, arguing that the Bible did not teach these false views.

In all of these examples, we see the same fundamental theme emerge. We've been here before. Historically, Christianity made claims which were at odds with the best science of the day. Details of the Biblical record of early history were particularly contentious. A minority sought to construct an alternative cosmology from scant Biblical references. Objections could be raised to a number of Biblical passages which could be read as assuming incorrect scientific ideas. We do not face an unprecedented situation; we face a situation much like that of the historical Church.

The situation has nevertheless changed. Many of the conflicts with classical science are now moot because classical science has been replaced by modern science. On many issues, modern science is more closely aligned with the Biblical view than classical science was. On other issues, better Biblical translations have helped eliminate conflicts. Conflict certainly remains, however. Biblical historical claims about early history remain contentious. It also remains possible to argue that the Bible assumes or teaches various scientific errors.

This history obviously contradicts a naive narrative in which scientific objections to Christianity only arose after the advent of modern science. It is clear that heliocentrism and the Galileo affair were not the beginning of the interaction of Christianity with science. Rather, the Church had been interacting with science for a millennium before that. However, it also undermines other less naive but still incorrect narratives.

For example, it shows the incorrectness of the warfare narrative. The warfare narrative contends that Christianity is fundamentally incompatible with, and therefore at war with, science. However, we find instead that historical Christians generally had no problem with scientific theories. They rejected them only when they thought they contradicted Scripture.

It also undermines the opposite narrative, in which the Church has historically simply accepted the science of the day. Certainly, it is the case that the Church was willing to accept most of the best science of its day. However, the Church was still quite willing to reject scientific ideas when those ideas contradicted the Biblical text. We find, in fact, that historical authors were very willing to draw scientific conclusions from the Biblical text. They were never willing to simply discount the authority of the Scriptures to speak to any issue, even ones of science. Christians who defended the compatibility of classical science with the Scriptures always did so by arguing that, when understood correctly, the Scriptures did not conflict with the science.

It is key to appreciate that historical Christians were discerning in how they applied the Biblical text to scientific questions. Not all of them made the correct decisions. But as a whole, they discerned when the text really demanded that they reject the scientific claims of their day.

From this, we can learn an important lesson: take heart. Scientific challenges to Scripture and our faith are not unprecedented. Rather, they have been with us from the early Church. We have always had to deal with arguments that the wisest ideas of the age show our faith to be in error. There has also always been disagreement within the Church about how to deal with some of these issues. Nonetheless, historically the Church has always torn down every stronghold that set itself against our faith, and there is every reason to believe it will continue to do so.

Historical Authors Accepted Biblical Authority

Many modern Christians reject the applicability or authority of the Scriptures to scientific questions. This is often expressed as the dictum, "the Bible is not a science textbook." Some will reinterpret Biblical passages which appear to conflict with scientific theories, often taking them metaphorically, symbolically, or phenomenologically. Others will argue that the Bible contains various scientific errors and that we should only accept the Bible's authority in more strictly religious matters. In either case, these Christians would never reject a scientific theory because it conflicts with the Biblical text.

Historical authors did not have this view; instead, they were quite willing to draw conclusions about strictly scientific questions from the Biblical text. This can be seen on issues where historical Christians rejected the science of their day on the basis of the Biblical text. For example, they rejected the idea that the heavens were eternally immutable, that matter had to have existed eternally, and that mankind was eternal. Some authors, such as Aquinas, went further in accommodating aspects of these ideas; nevertheless, even he rejected philosophical arguments that conflicted with the Biblical claims.

It can also be seen in cases where a minority of Christians advocated a flat-earth cosmology. They rejected the idea that gravity acted towards the center of the earth, insisted that the heavens were not a rotating sphere, and maintained that the earth was floating on the waters. For some or all of these points, these authors argued that the Scriptures taught their preferred cosmology. Their exegesis was flawed but the point remains that they thought the Bible certainly could speak to such questions.

Other Christians who defended classical science on these points did not merely dismiss the idea that the Bible could speak to such questions. Some, such as Augustine, did show some impatience or annoyance at such questions. Nevertheless, they consistently argued that the Scriptures, rightly interpreted, did not make the alleged claims. They were clear that *if* the Scriptures really taught a particular scientific claim, they were bound to believe it. Thus, they argued that the Scriptures did not teach these false scientific claims.

It can also be seen in the way that historical Christians handled scientific objections to the Biblical text. When dealing with issues such as the Bible's statements about the size of astronomical bodies, the source of the moon's light, and the pillars of the earth, historical Christians did not merely dismiss the idea that these statements could be making scientific claims. Rather, they argued that the text did not mean what the objector alleged the text to mean.

An easy approach to texts that appear to make claims at odds with scientific theories is to interpret those texts allegorically or symbolically. If the waters above the heavens are just a symbol or an allegory for a spiritual truth, then they do not pose any difficulty. In that case, we would not have to deal with the thorny scientific questions raised by the Biblical references to those waters. Indeed, the same strategy can be used to avoid any scientific difficulties found in the Biblical text. Any Biblical reference or claim, no matter how direct, can be simply dismissed by reading it as an allegory.

Certainly, the Bible does use allegory and symbolism; notable examples include prophetic visions and Jesus' parables. Ordinarily, we interpret texts allegorically when the context makes it clear that a symbolic interpretation is warranted. In almost all cases, it is pretty clear whether a particular passage is intended as a straightforward historical account or some sort of symbolic allegory. Nevertheless, a possible strategy to deal with scientific challenges is to take any passage posing such a challenge as an allegory, even if there is no contextual reason to think that it was intended as an allegory. This strategy has a certain effectiveness because, no matter the nature of the challenge, it can be resolved by taking it as an allegory.

Historically, this strategy has been rare. Origen does invoke this strategy. He is explicit in his book, *On First Principles*, that he thinks that sometimes the text is only true when understood allegorically and not when understood literally or according to the letter. However, other authors do not follow him on this point. Augustine, in some places, suggests allegorical interpretations but does not seem willing to simply leave the text with just an allegorical interpretation. He consistently tries to offer a literal interpretation of the text.

An important reason behind this is that historical Christians understood the difference between literal and allegorical interpretations of the text differently than we do. The modern Christian will tend to view it as a question of whether to understand a text literally or allegorically. The historical Christian would understand the text literally and allegorically at the same time. For example, if they read the story of David's defeating Goliath, they might interpret it both as a literal historical event that took place as well as an allegory of Christ's defeating Satan. Today, if someone presents an allegorical interpretation of a passage, most Christians would think of it as an alternative interpretation to a more literal one. However, the historical Christian was more likely to understand it as an additional, supplementary interpretation. Consequently, presenting an allegorical interpretation of a passage would not historically have seemed to resolve any scientific issues. Even if the allegorical meaning were accepted, the text would still have its literal meaning and the literal meaning would still pose a scientific problem.

This does not mean that historical Christians always read the text in a strictly literal fashion. They will read the text as engaging in metaphor, figures of speech, and hyperbole. Sometimes they offer interpretations that seem very awkward and unlikely. Nevertheless, we do not find historical authors arbitrarily reading texts as allegories in order to avoid a scientific difficulty.

Historical Christians did not either entirely reject or entirely accept the science of their day. Instead, they showed discernment, sometimes insisting that the Scriptures had to be trusted and the science rejected. In other cases, they argued that the Scriptures, rightly interpreted, were not in conflict with the science of the day.

When it came to questions about the eternity of the world, they were insistent that God had created the world from nothing in the recent past. Whether dealing with the immutability of the heavens, the conservation of matter, or the eternality of mankind, they always insisted that the world had not always existed, and philosophical arguments to the contrary were incorrect. For some cosmological questions, they also rejected the science of the day, insisting on the existence of a supercelestial realm, the solidity in some sense of the heavens, and the waters above the heavens. However, for many other cosmological questions, they were quite willing to accept the science of the day, accepting the classical theory of gravity, the rotating spherical heavens, notions of the nature of matter, the waters' being above the land, the relative sizes of the heavenly bodies, and the moon's not being the source of its own light. They defended the historical claims of the early chapters of Genesis, insisting that the sun really was created in the middle of Creation Week, that birds were created out of water, that the four rivers flowing from Eden still had a single source, that people living before the flood lived extended lives, and that the Ark was large enough to hold all the animals. However, in other cases, they were willing to accept the science of the day, arguing that it was compatible with the text. They argued that the waters' being gathered together were not incompatible with the many lakes and seas, that the description of a single spring watering the whole earth was not incompatible with many springs watering the earth, and that Methuselah did not actually survive the flood outside of the Ark.

The situation was somewhat more complicated than just described. Indeed, in many cases, there were minorities among historical authors who disagreed with the majority on whether to accept or reject a particular scientific claim. Notably, some people rejected the rotating spherical heavens, insisted the land floated on the waters, denied the existence of a supercelestial realm, or held that Methuselah had survived the flood outside of the Ark. Even so, the point remains that they were attempting to discern what Scripturally they had to believe, and what aspects of the science of their day they should have accepted.

This is not to say that they were correct in all of these judgments. Almost any Christian today would acknowledge that some of these judgments calls were incorrect. Few would insist on the solidity of the heavens, birds' being created out of water, that the four rivers still have a common source, or that a single spring watered the whole earth. However, the point is not the particular calls that they made based on the information that they had. The point, rather, is that they made these calls, seeking to determine what exactly the Scriptures taught on these questions.

When faced with a situation where the Scriptures appear to be at odds with a scientific theory, there are three possibilities. Firstly, the Scriptures themselves may be in error. Secondly, the scientific theory may be in error. Thirdly, the Scriptures may have been misunderstood so that there is no actual conflict. The historical authors never adopted the first approach; they never allowed that the Scriptures might be in error. They would sometimes reject scientific theories and sometimes argue that the Scriptures have been interpreted incorrectly, but they accepted the authority of the Scriptures. The modern Christian may choose to disagree with the historical Christian. They may insist that the Bible is only reliable on questions of science and faith. Or they may be willing to interpret the Bible in ways that a historical Christian would not have accepted. But they must acknowledge that, in doing so, they are departing from the historical practice of the faith.

Go Back to the Source

There is a common pattern. Problems are solved by going back to the source and reevaluating what we think based on a reconsideration of the Biblical text and the scientific data. Numerous problems are caused by not going back to the source.

Historical authors interpreted the text as teaching some scientific ideas of their own day. Origen tried to argue that the Scriptural and apocryphal texts referred to formless matter as the underlying material of all things. Several authors made a big deal about the Bible's accuracy in describing the sun and moon as great lights. Calvin took the Bible to teach that the waters being kept above the land were an indication of God's power. All of these ideas were the standard scientific views of the day. These authors tried to use the presence of these ideas as evidence for the inspiration of Scripture. But science has since changed. We no longer follow the notion of formless matter as the material of all things. The sun and moon are large, but they are dwarfed by many of the seemingly small stars in the sky. The sea is indeed below the land, not above it. Interpreting the text as teaching current scientific theories has repeatedly backfired as those theories changed.

Fundamentally, the problem was that they did not go back to the source. They were not interpreting these texts as they would have been understood by the original authors. These authors would have had no concept of the classical science notion of formless matter. The author of the creation account probably had no notion of how large either the sun or the moon was. The idea that the waters were above the land would have seemed bizarre and nonsensical to the Biblical authors. If historical Christians had gone back to the source and sought to understand what the original authors were referring to, they ought to have realized that there was no way they were referring to ideas that would not arise until much later.

Nevertheless, we should not judge these historical Christians for failing on this point. Reading Biblical texts in light of our own scientific ideas is hard to avoid. It is very easy to inadvertently read the text as referring to some modern idea without realizing it. Some may not have even realized how specific to their own time a particular idea was. Even when we realize that the idea is modern, it can be very tempting to read it into the text. But history shows that we should resist that temptation.

A variation on this is found among those who rejected the scientific ideas of their day. This includes those who defended an earth floating on waters, rejected a rotating heavenly sphere, or argued that the earth had definite ends. Some authors appealed to the Scriptures to defend their ideas on these questions, rejecting the scientific consensus of their day. These authors would turn out to be in error. The earth does not float on water; the earth itself rotates, which is effectively equivalent to the heavens' rotating; and the earth is spherical and does not have ends.

But they again made the same mistake; they did not go back to the source. They read the Bible as commenting on the scientific debates of their own day. The Biblical authors would have never thought about whether the earth could be floating on the waters, if the heavens were some sort of rotating sphere, or if there was an actual end of the earth. These ideas only arise with at least a primitive cosmology, and the Biblical authors were too primitive to even have an incorrect cosmology.

Many issues arose due to translation errors in the Septuagint. According to the Septuagint, the heavens were some sort of solid firmament. The earth was created invisible and possibly out of formless matter. Birds were created from water. A single spring was responsible for watering the whole earth. Methuselah survived the flood despite not being on the Ark. The heavens are compared to a skin, a vault, and smoke. Partridges steal the eggs of other birds. Additionally, the Septuagint inconsistently translated or transliterated Hebrew names. All of these errors led to confusion and difficulty for the early Christians, who were only familiar with the Septuagint and not the original Hebrew text.

But again the problem was not going back to the source. They did not go back to the original text but were relying on a flawed translation. Sometimes, such as in the case of Methuselah, they realized this and did go back to the source. Other cases were corrected as new translations were used, which avoided the errors of the Septuagint. Even today, there are cases where the Septuagint's translation still negatively influences modern translations. In all these cases, the problem is resolved by going back to the source.

Sometimes authors adopted very dubious exegeses in an attempt to align the Biblical text with a scientific theory. Augustine and Aquinas both try to argue that humid air is close enough to water to explain the Septuagint's claim that birds were made out of water. Calvin has great difficulty in trying to explain why the Bible says that the waters are below the earth when he thinks they are at a higher elevation. Augustine tries to argue that numerous springs all over the earth could be referred to a single spring. They were highly dubious ways to read these texts.

These problems were resolved by going back to the source. In some cases, these were simply translation errors, and going back to the source resolves the problem much better than awkward exegesis. In other cases, the scientific theory was simply wrong and thus there was no need to reconcile it with the Biblical text. In that case, we had to go back to the scientific source, re-evaluating what the scientific data meant. In still other cases, there were simply better and simpler ways to understand the text without awkward exegesis.

Again and again, going back to the source solves scientific issues. Historically and today, the solution is to go back to the source. This means going back to the Scriptures and discerning what they actually intend to say. It means going back to the scientific data and evaluating what it actually shows. So, take heart, follow the authority of Scripture, and go back to the source.

Bibliography

Ambrose. *Commentary of Saint Ambrose on the Gospel According to Saint Luke*. Translated by Ide M. Ni Riain. Dublin, Ireland: Haleyon Press, 2001.

Ambrose of Milan. "Hexameron." In *Hexameron, Paradise, and Cain and Abel*, translated by John J. Savage, Vol. 42. The Fathers of the Church: A New Translation, n.d.

Aquinas, Thomas. *The Summa Theologiæ of St. Thomas Aquinas*. Second and Revised Edition., 1920. https://www.newadvent.org/summa/.

Arnobius. "Against the Heathen." In *Ante-Nicene Fathers*, edited by Alexander Roberts, James Donaldson, and A. Cleveland Coxe, translated by Hamilton Bryce and Hugh Campbell, Vol. 6. Buffalo, NY: Christian Literature Publishing Co., 1886.

Athanasius. "On the Incarnation of the Word." In *Nicene and Post-Nicene Fathers, Second Series*, edited by Philip Schaff and Henry Wace, translated by Archibald Robertson, Vol. 4. Buffalo, NY: Christian Literature Publishing Co., 1892.

Athanasius of Alexandria. "Against the Heathen." In *Nicene and Post-Nicene Fathers, Second Series*, edited by Philip Schaff and Henry Wace, translated by Archibald Robertson, Vol. 4. Buffalo, NY: Christian Literature Publishing Co., 1892.

Athenagoras. "A Plea for the Christians." In *Ante-Nicene Fathers*, edited by Alexander Roberts, James Donaldson, and A. Cleveland Coxe, translated by B.P. Pratten, Vol. 2. Buffalo, NY: Christian Literature Publishing Co., 1885.

Augustine. "Confessions." In *Nicene and Post-Nicene Fathers, First Series*, edited by Philip Schaff, translated by J.G. Pilkington, Vol. 1. Buffalo, NY: Christian Literature Publishing Co., 1887.

———. "Contra Faustum." In *Nicene and Post-Nicene Fathers, First Series*, edited by Philip Schaff, translated by Richard Stothert, Vol. 4. Buffalo, NY: Christian Literature Publishing Co., 1887.

———. "Exposition on Psalm 55." In *Nicene and Post-Nicene Fathers, First Series*, edited by Philip Schaff, translated by J.E. Tweed, Vol. 8. Buffalo, NY: Christian Literature Publishing Co., 1888.

———. "On The Grace of Christ, and on Original Sin." In *Nicene and Post-Nicene Fathers, First Series*, translated by Peter Holmes, Robert Ernest Wallis, and Benjamin B. Warfield, Vol. 5. Buffalo, NY: Christian Literature Publishing Co., 1887.

———. "On the Morals of the Manichaeans." In *Nicene and Post-Nicene Fathers, First Series*, edited by Philip Schaff, translated by Richard Stothert, Vol. 4. Buffalo, NY: Christian Literature Publishing Co., 1887.

———. "Questions on Heptatechum." In *The Works of Saint Augustine. Vol. 14: Pt. 1. Books Writings on the Old Testament*, Vol. 14. Brooklyn, NY: New City Pr, 2016.

———. *Sermons 184-229Z, on the Liturgical Seasons*. Edited by John E. Rotelle. Translated by Edmund Hill. The Works of Saint Augustine 6. New Rochelle (N.Y.): New City press, 1993. 252.009 01.

———. "The City of God." In *Nicene and Post-Nicene Fathers, First Series*, edited by Philip Schaff, translated by Marcus Dods, Vol. 2. Buffalo, NY, 1887.

———. *The Literal Meaning of Genesis*. Translated by John H. Taylor. Ancient Christian Writers 41. New York, NY: Paulist Pr, 1982.

———. "Two Books on Genesis against the Manichees." In *On Genesis: Two Books on Genesis against the Manichees and On the Literal Interpretation of Genesis: An Unfinished Book (The Fathers of the Church, Volume 84)*, translated by Roland J. Teske. Catholic University of America Press, 2010.

Basil of Caesarea. "Hexaemeron." In *Nicene and Post-Nicene Fathers, Second Series*, edited by Schaff, Philip and Wace, Henry, translated by Blomfield Jackson. Buffalo, NY: Christian Literature Publishing Co, 1885.

Bede. "Commentary on Genesis." In *Commentaries on Genesis 1-3*, edited by Michael Glerup, translated by Robert C. Hill and Carmen Hardin. Ancient Christian Texts. Downers Grove, Ill.: IVP Academic, c2010, 2010. BS1235.53 .C655 2010.

Buridan, John, and Marshall Clagett. "Questions on the Eight Books of the Physics of Aristotle." In *The Science of Mechanics in the Middle Ages*. Madison: University of Wisconsin press, 1979.

Cajori, Florian. "The Age of the Sun and the Earth." Scientific American. Accessed February 14, 2024. https://www.scientificamerican.com/article/the-age-of-the-sun-and-the-earth/.

Calvin, John. "A Warning against Judiciary Astrology and Other Prevalent Curiosities," n.d., 157–90.

———. *Calvin's Commentary on the Bible*, n.d.

Chambers, Nathan J. *Reconsidering Creation Ex Nihilo in Genesis 1*. Vol. 19. Penn State University Press, 2020. https://doi.org/10.5325/j.ctv1hcg09r.

Chrysostom, John. *Homilies on Genesis*. The Fathers of the Church, a New Translation, v. 74. Washington, D.C: Catholic University of America Press, 1999. BR60.F3 BS1235 J616 1999.

———. "Homilies on Hebrews." In *Nicene and Post-Nicene Fathers, First Series*, edited by Philip Schaff, translated by Frederic Gradiner, Vol. 14. Buffalo, NY: Christian Literature Publishing Co., 1889.

———. "Homilies on Matthew." In *Nicene and Post-Nicene Fathers, First Series*, edited by Philip Schaff, translated by George Prevost and M. B. Riddle, Vol. 10. Buffalo, NY: Christian Literature Publishing Co., 1888.

———. "Homilies on Romans." In *Nicene and Post-Nicene Fathers, First Series*, translated by J. Walker, J. Sheppard, H. Browne, and George B. Stevens, Vol. 11. Buffalo, NY: Christian Literature Publishing Co, 1889.

———. "Homilies on the Statues." In *Nicene and Post-Nicene Fathers, First Series*, edited by Philip Schaff, translated by W.R.W. Stephens, Vol. 9. Buffalo, NY: Christian Literature Publishing Co., 1889.

Copan, Paul. "Creation Ex Nihilo or Ex Materia? A Critique of the Mormon Doctrine of Creation." *Southern Baptist Journal of Theology* 9, no. 2 (2005): 32–49.

Copan, Paul, and William Lane Craig. *Creation out of Nothing: A Biblical, Philosophical, and Scientific Exploration*. Grand Rapids (Mich.): Baker Academic, 2004.

Craig, William Lane, and Quentin Smith. *Theism, Atheism, and Big Bang Cosmology*. Oxford: Clarendon Press, 2003.

Dante Alighieri. *A Question of the Water and of the Land*. Translated by Charles Hamilton Bromby. London: Ballantyne Press, 1897.

———. *De Vulgari Eloquentia*. Translated by Steven Botterill, n.d.

———. *The Banquet*. Translated by Elizabeth Price Sayer. Project Gutenberg, 2004.

———. *The Divine Comedy*. Translated by H. F. Cary, n.d.

Ephrem the Syrian. *Commentary on Genesis*, n.d. http://www2. iath.virginia.edu/anderson/commentaries/EphGen.html.

Eusebius. *Chronicle*. Translated by Robert Bedrosian, n.d. https://archive.org/details/EusebiusChroniclechronicon /page/n1/mode/2up.

Gregory of Nyssa. "Answer to Eunomius' Second Book." In *Nicene and Post-Nicene Fathers, Second Series*, edited by Philip Schaff and Henry Wace, translated by M. Day, Vol. 5. Buffalo, NY: Christian Literature Publishing Co., 1893.

———. "On the Soul and the Resurrection." In *Nicene and Post-Nicene Fathers, Second Series*, edited by Philip Schaff and Henry Wace, translated by William Moore and Henry Austin Wilson, Vol. 5. Buffalo, NY: Christian Literature Publishing Co., 1893.

Hart, Pauly. *Biblical Cosmology*, n.d.

Hilary of Poitiers. "On the Councils." In *Nicene and Post-Nicene Fathers, Second Series*, edited by Philip Schaff, translated by E.W. Watson and L. Pullan, Vol. 9. Buffalo, NY: Christian Literature Publishing Co., 1899.

Hippolytus. "On Christ and Antichrist." In *Ante-Nicene Fathers*, edited by Alexander Roberts, James Donaldson, and A. Cleveland Coxe, translated by J.H. MacMahon, Vol. 5. Buffalo, NY: Christian Literature Publishing Co., 1886.

———. "The Refutation of All Heresies." In *Ante-Nicene Fathers*, edited by Alexander Roberts, James Donaldson, and

A. Cleveland Coxe, translated by J.H. MacMahon, Vol. 5. Buffalo, NY: Christian Literature Publishing Co., 1886.

Hu, Wangjie, Ziqian Hao, Pengyuan Du, Fabio Di Vincenzo, Giorgio Manzi, Jialong Cui, Yun-Xin Fu, Yi-Hsuan Pan, and Haipeng Li. "Genomic Inference of a Severe Human Bottleneck during the Early to Middle Pleistocene Transition." *Science* 381, no. 6661 (September 2023): 979–84. https://doi.org/10.1126/science.abq7487.

Huff, Chad D., Jinchuan Xing, Alan R. Rogers, David Witherspoon, and Lynn B. Jorde. "Mobile Elements Reveal Small Population Size in the Ancient Ancestors of Homo Sapiens." *Proceedings of the National Academy of Sciences of the United States of America* 107, no. 5 (February 2, 2010): 2147–52. https://doi.org/10.1073/pnas.0909000107.

Indicopleustes, Cosmas. *Christian Topography*. Translated by John Watson McCrindle, n.d.

———. *Christian Topography*. Translated by John Watson McCrindle, n.d.

Jerome. *Commentary on Ecclesiastes*. Edited by John Litteral. Translated by Robin McGregor. Ancient Bible Commentaries in English. West Monroe, La: Consolamini Publications, 2015.

———. *Commentary on Jeremiah*. Edited by Christopher A. Hall. Translated by Michael Graves. Ancient Christian Texts. Downers Grove, Ill: IVP Academic, 2011.

———. "Letter 58." In *Nicene and Post-Nicene Fathers, Second Series*, edited by Philip Schaff and Henry Wace, translated by W.H. Freemantle, G. Lewis, and W.G. Martley, Vol. 6. Buffalo, NY: Christian Literature Publishing Co., 1893.

Jerome, and Robert Hayward. *Saint Jerome's Hebrew Questions on Genesis*. Oxford Early Christian Studies. Oxford, England : New York: Clarendon Press ; Oxford University Press, 1995. BS1235 .J4713 1995.

John of Damascus. "An Exposition of the Orthodox Faith." In *Nicene and Post-Nicene Fathers, Second Series*, edited by Philip Schaff and Henry Wace, translated by E.W. Watson and L. Pullan, Vol. 9. Buffalo, NY: Christian Literature Publishing Co., 1899.

Johnson, Phillip E. *The Right Questions: Truth, Meaning & Public Debate*. Downers Grove, Ill: InterVarsity Press, 2002.

Lactantius. "The Divine Institutes." In *Ante-Nicene Fathers*, edited by Alexander Roberts, James Donaldson, and A. Cleveland Coxe, translated by William Fletcher, Vol. 7. Buffalo, NY: Christian Literature Publishing Co., n.d.

Leo the Great. "Sermons of St Leo The Great." In *Nicene and Post-Nicene Fathers, Second Series*, edited by Philip Schaff and Henry Wace, translated by Charles Lett Feltoe, Vol. 12. Buffalo, NY: Christian Literature Publishing Co., 1895.

Lewis, C. S. *Surprised by Joy: The Shape of My Early Life*. A Harvest Book. New York: Harcourt, Brace & World, 1984.

Luther, Martin. *Luther On The Creation: A Critical and Devotional Commentry on Genesis*. Translated by Henry Coles. Vol. 1. The Precious And Sacred Writings Of Martin Luther. Minneapolis, Minnesota, USA: Lutherans in all Lands co., 1904.

Luther, Martin, and C. M. Jacobs. "A Treatise on Usury." In *Works Of Martin Luther*, Vol. IV. Philadelphia, Pennsylvania: A. J. Holman Company and The Castle Press, 1931.

Martyr, Justin. "The First Apology." In *Ante-Nicene Fathers*, edited by Alexander Roberts, James Donaldson, and A. Cleveland Coxe, translated by Marcus Dods and George Reith, Vol. 1. Buffalo, NY: Christian Literature Publishing Co., 1885.

Melanchthon, Philip. "Preface." In *On the Judgments of Nativities: Three Books*, by Johannes Schoener, translated by Robert Hand, n.d.

461

Moore, Robert A. "The Impossible Voyage of Noah's Ark." Creation/Evolution Magazine, P, 0 1983.

Origen. "Commentary on the Gospel of John." In *Ante-Nicene Fathers*, edited and translated by Allan Menzies, Vol. 9. Buffalo, NY: Christian Literature Publishing Co., 1896.

———. "Contra Celsum." In *Ante-Nicean Fathers*, edited by Roberts, Alexander, Donaldson, James, and Coxe, A. Cleveland, translated by Frederick Crombie, Vol. 4. Buffalo, NY: Christian Literature Publishing Co, 1885.

———. *Homilies on Genesis and Exodus*. Translated by Ronald E. Heine. Catholic University of America Press, 1982. https://doi.org/10.2307/j.ctt32b3pv.

———. *Homilies on Jeremiah and I Kings 28*. Translated by John Clark Smith. Vol. 97. The Fathers of the Church. Catholic University of America, 1998.

———. *The Philocalia of Origen*. Translated by Rev. George Lewis, 1911.

Origen, John C. Cavadini, and Henri de Lubac. *On First Principles*. Notre Dame, Indiana: Christian Classics, 2013. BR65.O568 E56 2013.

Origen, and Thomas P. Scheck. *Homilies on Numbers*. Edited by Christopher A. Hall. Ancient Christian Texts. Downers Grove, Ill: IVP Academic, 2009.

Ostler, Blake T. "Out of Nothing: A History of Creation *Ex Nihilo* in Early Christian Thought." *The FARMS Review* 17 (2005), no. 2 (January 1, 2005): 253–320. https://doi.org/10.5406/farmsreview.17.2.0253.

Philoponus, John. *Against Aristotle on the Eternity of the World*. Translated by Christian Wildberg. The Ancient Commentators on Aristotle. London: G. Duckworth, 1987. 1.

Plato. "Phaedrus." In *The Dialogues of Plato*, translated by David Horan, 2021.

———. *Timaeus*. Translated by Benjamin Jowett, n.d.

Pliny. *The Natural History*, n.d.

Plutarch. "Of the Tranquillity and Contentment of Mind." In *Plutarch's Moral Essays*, edited by Ernest Rhys, translated by Philemon Holland. Everyman's Library, 1911.

Remigius of Auxerre. "Exposition on Genesis." In *The Book of Genesis*, edited by Joy A. Schroeder. The Bible in Medieval Tradition. Grand Rapids, Michigan ; Cambridge, U.K: William B. Eerdmans Publishing Company, 2015. BS1235.52 .B658 2015.

Severian. "Homilies on Creation and Fall." In *Commentaries on Genesis 1-3*, edited by Michael Glerup, translated by Robert C. Hill and Carmen Hardin. Ancient Christian Texts. Downers Grove, Ill. : IVP Academic, c2010, 2010. BS1235.53 .C655 2010.

Sungenis, Robert, and Bennet, Robert. *Galileo Was Wrong: The Church Was Right: The Evidence from Modern Science*. 12th Edition. Vol. 1, 2022.

Tatian. "Tatian's Address to the Greeks." In *Ante-Nicene Fathers*, edited by Alexander Roberts, James Donaldson, and A. Cleveland Coxe, translated by J.E. Ryland, Vol. 2. Buffalo, NY: Christian Literature Publishing Co., 1885., 1885.

Tertullian. "Against Hermogenes." In *Ante-Nicene Fathers*, edited by Alexander Roberts, James Donaldson, and A. Cleveland Coxe, translated by Peter Holmes, Vol. 3. Buffalo, NY: Christian Literature Publishing Co., 1885.

———. "On Idolatry." In *Ante-Nicene Fathers*, edited by Alexander Roberts, James Donaldson, and A. Cleveland Coxe, translated by S. Thelwall, Vol. 3. Buffalo, NY: Christian Literature Publishing Co., 1885.

"The Garden of Eden." Accessed February 6, 2024. https://www.asa3.org/ASA/PSCF/2000/PSCF3-00Hill.html.

Theophilus of Antioch. "To Autolycus." In *Ante-Nicene Fathers*, edited by Alexander Roberts, James Donaldson, and A.

Cleveland Coxe, translated by Marcus Dods, Vol. 2. Buffalo, NY: Christian Literature Publishing Co., 1885.

Wildberg, Christian. "John Philoponus." In *The Stanford Encyclopedia of Philosophy*, edited by Edward N. Zalta, Winter 2021. Metaphysics Research Lab, Stanford University, 2021. https://plato.stanford.edu/archives/win2021/entries/philoponus/.

Woodmorappe, John. *Noah's Ark: A Feasibility Study*. Santee, Calif: Institute for Creation Research, 1996. BS658 .W64 1996.

Index

Index

Index

Index

www.ingramcontent.com/pod-product-compliance
Lightning Source LLC
Chambersburg PA
CBHW060123130626
46556CB00006B/2210